Lecture Notes in Mathematics

Edited by A. Dold and B. Eckmann

769

Jörg Flum
Martin Ziegler

Topological Model Theory

Springer-Verlag
Berlin Heidelberg New York 1980

Authors

Jörg Flum
Mathematisches Institut
Abt. für math. Logik
Universität Freiburg
D-7800 Freiburg

Martin Ziegler
Mathematisches Institut
Beringstr. 4
D-5300 Bonn

AMS Subject Classifications (1980): 03B60, 03C90, 03D35, 12L99, 20A15, 46A99, 54-02

ISBN 3-540-09732-5 Springer-Verlag Berlin Heidelberg New York
ISBN 0-387-09732-5 Springer-Verlag New York Heidelberg Berlin

Library of Congress Cataloging in Publication Data
Flum, Jörg.
Topological model theory.
(Lecture notes in mathematics; 769)
Bibliography: p.
Includes index.
1. Topological spaces. 2. Model theory. I. Ziegler, Martin, joint author. II. Title.
III. Series: Lecture notes in mathematics (Berlin); 769.
QA3.L28 no. 769 [QA611.3] 510s [515.7'3] 79-29724
ISBN 0-387-09732-5

Printing and binding: Beltz Offsetdruck, Hemsbach/Bergstr.
2141/3140-543210

To

Siegrid and Gisela

TABLE OF CONTENTS

INTRODUCTION

The task of model theory is to investigate mathematical structures with the aid of formal languages. Classical model theory deals with algebraic structures. Topological model theory investigates topological structures. A topological structure is a pair $(\mathfrak{A}, \sigma)$ consisting of an algebraic structure $\mathfrak{A}$ and a topology σ on A. Topological groups and topological vector spaces are examples. The formal language in the study of topological structures is L_t. This is the fragment of the (monadic) second-order language (the set variables ranging over the topology σ) obtained by allowing quantification over set variables in the form $\exists X(t \in X \wedge \varphi)$, where t is a term and the second-order variable X occurs only negatively in φ (and dually for the universal quantifier). Intuitively, L_t allows only quantifications over sufficiently small neighborhoods of a point.

The reasons for the distinguished role that L_t plays in topological model theory are twofold. On one hand, many topological notions are expressible in L_t, e.g. most of the freshman calculus formulas as "continuity"

$$\forall x \, \forall Y(fx \in Y \rightarrow \exists X(x \in X \wedge \forall z(z \in X \rightarrow fz \in Y))).$$

On the other hand, the expressive power is not too strong, so that a great deal of classical model theory generalizes to L_t. For example, L_t satisfies a compactness theorem and a Löwenheim-Skolem theorem. In fact, L_t is a maximal logic with these properties ("Lindstöm theorem").

While in the second part we study concrete L_t-theories, the first part contains general model-theoretic results. The exposition shows that it is possible to give a parallel treatment of classical and topological theory, since in many cases the results of topological model theory are obtained using refinements of classical methods. On the other hand there are many new problems which have no classical counterpart.

The content of the sections is the following.

§ 1 contains preliminaries. While second-order language is too rich to allow a fruitful model theory, central theorems of classical model theory remain true if we restrict to invariant second-order formulas. Here φ is called invariant, if for all topological structures $(\mathfrak{A},\sigma)$,

$$(\mathfrak{A},\sigma) \models \varphi \quad \text{iff} \quad (\mathfrak{A},\tau) \models \varphi \text{ holds for all bases } \tau \text{ of } \sigma.$$

Many topological notions are invariant; e.g. "Hausdorff", since when checking the Hausdorff property it suffices to look at the open sets of a basis.

In section 2 we introduce the language L_t; L_t-formulas are invariant, later on (§ 4) we show the converse: each invariant formula is equivalent to an L_t-formula. -

In section 3 we derive for L_t some results (compactness theorem, Löwenheim-Skolem theorem,...) which follow immediately from the fact that L_t may be viewed as a two-sorted first-order language.

We generalize in section 4 the Ehrenfeucht-Fraïssé characterization of elementary equivalence and the Keisler-Shelah ultrapower theorem. For this we introduce for topological structures back and forth methods, which also will be an important tool later on. In § 5 we prove the L_t interpolation theorem, and derive preservation theorems for some relations between topological structures. In particular, we characterize the sentences which are preserved by dense or open substructures. In § 6 we show that operations like the product and sum operation on topological structures preserve L_t-equivalence.

Section 7 contains the L_t-definability theory. Besides the problem of the explicit definability of relations, which in classical model theory are solved by the theorems of Beth, Svenonius,... , there arises in topological model theory also the problem of the explicit definability of a topology.

In § 8 we first prove a Lindström-type characterization of L_t. - There are natural languages for several other classes of second-order structures like structures on uniform spaces, structures on proximity spaces. All these languages as well as L_t can be interpreted in the language L_m for monotone structures.

The omitting types theorem fails for L_t; we show this in section 9, where we also prove on omitting types theorem for a fragment of L_t, which will be useful in the second part. The last section is devoted to the infinitary language $(L_{\omega_1\omega})_t$. We generalize many results to this language showing that each invariant Σ_1-sentence over $(L_{\omega_1\omega})_2$ is equivalent in countable topological structures to a game sentence, whose countable approximations are in $(L_{\omega_1\omega})_t$.- We remark that some results like Scott's isomorphism theorem do not generalize to $(L_{\omega_1\omega})_t$.

The second part can be read without the complete knowledge of the first part. Essentially only §§ 1 - 4 are presupposed. The content of the sections of the second part is the following:

§ 1 Topological spaces.

We investigate decidability of some theories and determine their (L_t-) elementary types. For many classes of spaces, which do not share strong separation properties like T_3, the (L_t-)theory turns out to be undecidable. For T_3-spaces not only a decision procedure is given, but also a complete description of their elementary types by certain invariants. As a byproduct we get simple characterizations of the finitely axiomatized and of the $\aleph_0$-categorical T_3-spaces.

§ 2 Topological abelian groups.

Three theorems are proved:

1) The theory of all Hausdorff topological abelian groups is hereditarily undecidable.

2) The theory of torsionfree topological abelian groups with continuous (partial) division by all natural numbers is decidable.

3) The theory of all topological abelian groups A for which nA is closed and division by n is continuous is decidable.

§ 3 Topological fields.

We describe the L_t-elementary class of locally bounded topological fields (and other related classes) as class of structures which are L_t-equivalent to a topological field, where the filter of neighborhoods of zero is generated by the non-zero ideals of a proper local subring of K having K as quotient field.

V-topologies correspond to valuation rings. This fact has some applications in the theory of V-topological fields.- Finally we give L_t-axiomatizations of the topological fields $\mathbb{R}$ and $\mathbb{C}$.

§ 4 Topological vector spaces.

We give a simple axiomatization of the L_t-theory of the class of locally bounded real topological vector spaces. If we fix the dimension, then this theory is complete.

The L_t-elementary type of a locally bounded real topological vector space V with a distinguished subspace H is determined by the dimensions of H, $\bar{H}/_H$ and $V/_{\bar{H}}$ (where $\bar{H}$ denotes the closure of H). As an application we show that the L_t-theory of surjective and continuous linear mappings (essentially) can be axiomatized by the open mapping theorem.- Finally we determine the L_t-elementary properties of structures (V,V',[,]), where V is a real normed space, V' its dual space and [,] the canonical bilinear form.

The present book arose fom a course in topological model theory given by the second author at the University of Freiburg during the summer of 1977. We have collected all references and historical remarks on the results in the text in separate sections at the end of the first and the second part.

§ 1 Preliminaries

We denote similarity types by L,L',... . They are sets of predicate symbols (P,Q,R,...) and function symbols (f,g,...). Sometimes 0-placed function symbols are called constants and denoted by c,d,... . - $(\mathfrak{A},\sigma)$ is called a <u>weak L-structure</u> if $\mathfrak{A}$ is an L-structure in the usual sense and σ is a non-empty subset of the power set P(A) of A. If σ is a topology on A, we call $(\mathfrak{A},\sigma)$ a <u>topological structure</u>.

By $L_{\omega\omega}$ we denote the first-order language associated with L. It is obtained by introducing (individual) variables $w_0,w_1,\ldots$, forming terms and atomic formulas as usual, closing under the logical operations of $\neg,\wedge,\vee,\exists$ and $\forall$. $\rightarrow$ and $\leftrightarrow$ will be regarded as abbreviations, x,y,... will denote variables. - The (monadic) second-order language L_2 is obtained from $L_{\omega\omega}$ by adding the symbol $\in$ and set variables $W_0,W_1,\ldots$ (denoted by X,Y,...). New atomic formulas $t \in X$, where t is a term of L, are allowed. A formation rule is added to those of $L_{\omega\omega}$:

If φ is a formula so are $\exists X\varphi$ and $\forall X\varphi$.

The meaning of a formula of L_2 in a weak structure $(\mathfrak{A},\sigma)$ is defined in the obvious way: quantified set variables range over σ. (Note that we did not introduce formulas of the form X = Y, however they are definable in L_2.)

For the sentence of L_2

$$\varphi_{haus} = \forall x\, \forall y(\neg\, x = y \rightarrow \exists X\, \exists Y(x \in X \wedge y \in Y \wedge \forall z \neg\, (z \in X \wedge \in Y))) ,$$

and any topological structure $(\mathfrak{A},\sigma)$, we have

$$(\mathfrak{A},\sigma) \models \varphi_{haus} \quad \text{iff} \quad \sigma \text{ is a Hausdorff topology.}$$

Similarly the notions of a regular, a normal or a connected topology are expressible in L_2.

The logic L_2 (using weak structures as models) is reducible to a suitable (two sorted) first-order logic. Hence L_2 satisfies central model-theoretic theorems such as the compactness theorem, the completeness theorem and the Löwenheim-Skolem theorem, e.g.

1.1 <u>Compactness theorem</u>. A set of L_2-sentences has a weak model if every finite subset does.

This is not true if we restrict to topological structures as models: For

$$\varphi_{disc} = \forall x\, \exists X\, \forall y(y \in X \leftrightarrow y = x)\ ,$$

and any topological structure $(\mathfrak{A},\sigma)$, we have

$$(\mathfrak{A},\sigma) \models \varphi_{disc} \quad \text{iff} \quad \sigma \text{ is the discrete topology on } A$$
$$\text{iff} \quad \sigma = P(A).$$

Therefore, full monadic second-order logic is interpretable if we restrict to topological structures. Hence the compactness theorem, the completeness theorem and the Löwenheim-Skolem theorem do not longer hold. - In particular there is no $\varphi \in L_2$ such that

$$(\mathfrak{A},\sigma) \models \varphi \quad \text{iff} \quad \sigma \text{ is a topology}$$

holds for all weak structures $(\mathfrak{A},\sigma)$.

On the other hand to be the basis of a topology is expressible in L_2: Let

$$\varphi_{bas} = \forall x\, \exists X\, x \in X \wedge \forall x\, \forall X\, \forall Y(x \in X \wedge x \in Y \rightarrow$$
$$\exists Z(x \in Z \wedge \forall z(z \in Z \rightarrow (z \in X \wedge z \in Y)))).$$

Then

$$(\mathfrak{A},\sigma) \models \varphi_{bas} \quad \text{iff} \quad \sigma \text{ is basis of a topology on } A.$$

In the next section we will make use of this fact, when we introduce a sublanguage of L_2 which satisfies the basic modeltheoretic theorems even if we restrict to topological structures.

For $\sigma \subset P(A)$, $\sigma \neq \emptyset$, we denote by $\tilde{\sigma}$ the smallest subset of $P(A)$ containing σ and closed under unions,

$$\tilde{\sigma} = \{\bigcup s \mid s \subset \sigma\}.$$

Hence

$$(\mathfrak{A},\sigma) \models \varphi_{bas} \quad \text{iff} \quad \tilde{\sigma} \text{ is a topology.}$$

To prove that a function is continuous or that a topological space is Hausdorff, it suffices to test or to look at the open sets of a basis. These properties are "invariant for topologies" in the sense of the next definition.

1.2 Definition. Let φ be an L_2-sentence.

(i) φ is invariant if for all $(\mathfrak{A},\sigma)$:

$$(\mathfrak{A},\sigma) \models \varphi \quad \text{iff} \quad (\mathfrak{A},\tilde{\sigma}) \models \varphi.$$

(ii) φ is invariant for topologies if for all $(\mathfrak{A},\sigma)$ such that $\tilde{\sigma}$ is a topology,

$$(\mathfrak{A},\sigma) \models \varphi \quad \text{iff} \quad (\mathfrak{A},\tilde{\sigma}) \models \varphi.$$

Each invariant sentence is invariant for topologies. Note that φ is invariant for topologies if and only if for all topological structures $(\mathfrak{A},\tau)$ and any basis σ of τ one has

$$(\mathfrak{A},\sigma) \models \varphi \quad \text{iff} \quad (\mathfrak{A},\tau) \models \varphi.$$

Each sentence of the sublanguage L_t of L_2 that we introduce in the next section is invariant. Later on we will show the converse: Each invariant (invariant for topologies) L_2-sentence is equivalent (in topological structures) to an L_t-sentence.

1.3 Exercise. (a) Show that the notions "Hausdorff", "regular", "discrete" may be expressed by L_2-sentences that are invariant for toplogies.
(b) For unary $f \in L$, $\forall X \, \forall x(x \in X \to \exists Y(fx \in Y \wedge \forall y(y \in Y \to \exists z \in X \; fz = y)))$ is a sentence invariant for topologies expressing that f is an open map.
(c) For unary $P \in L$, $\exists X \, \forall y(y \in X \leftrightarrow Py)$ is a sentence not invariant for topologies. In topological structures it expresses that P is open (but see 2.5 (b)).
(d) Give an example of an L_2-sentence invariant for topologies that is not invariant.

1.4 Exercise. (Hintikka sets and term models). Suppose L is given. Let C be a countable set of new constants and $\mathfrak{U}$ a countable set of "set constants". Denote by $L(C,\mathfrak{U})_2$ the language defined as $(L \cup C)_2$ but using the additional atomic formulas $t \in U$ (for $U \in \mathfrak{U}$). Basic terms are the terms of the form $fc_1,\ldots c_n$ (with $c_1,\ldots,c_n \in C$) and the constants in C. Let Ω be a set of $L(C,\mathfrak{U})_2$-sentences in negation normal form (for a definition see the beginning of the next section). Ω is said to be a Hintikka set iff (i) - (x) hold:

(i) For each atomic φ of the form $c_1 = c_2$, $Rc_1 \ldots c_n$ or $c \in U$ (where $c_i, c \in C$ and $U \in \mathfrak{U}$) either $\varphi \notin \Omega$ or $\neg \varphi \notin \Omega$.

(ii) If $\varphi_1 \wedge \varphi_2 \in \Omega$ then $\varphi_1 \in \Omega$ and $\varphi_2 \in \Omega$.

(iii) If $\varphi_1 \vee \varphi_2 \in \Omega$ then $\varphi_1 \in \Omega$ or $\varphi_2 \in \Omega$.

(iv) If $\forall x\, \varphi \in \Omega$ then for all $c \in C$, $\varphi\frac{c}{x} \in \Omega$.

(v) If $\exists x\, \varphi \in \Omega$ then for some $c \in C$, $\varphi\frac{c}{x} \in \Omega$.

(vi) If $\forall X\, \varphi \in \Omega$ then for all $U \in \mathfrak{U}$, $\varphi\frac{U}{X} \in \Omega$.

(vii) If $\exists X\, \varphi \in \Omega$ then for some $U \in \mathfrak{U}$, $\varphi\frac{U}{X} \in \Omega$.

(viii) For all $c \in C$, $c = c \in \Omega$.

(ix) If t is a basic term, then for some $c \in C$, $t = c \in \Omega$.

(x) If φ is atomic or negated atomic and t is a basic term such that for some $c \in C$ and some variable x, $t = c \in \Omega$, and $\varphi\frac{t}{x} \in \Omega$, then $\varphi\frac{c}{x} \in \Omega$.

($\varphi\frac{t}{x}$ and similarly $\varphi\frac{U}{X}$, is obtained by replacing each free occurence of x in φ by t).

Suppose Ω is a Hintikka set. For $c_1, c_2 \in C$, let

$$c_1 \sim c_2 \quad \text{iff} \quad c_1 = c_2 \in \Omega.$$

Show that $\sim$ is an equivalence relation. Let $\bar{c}$ be the equivalence class of c. Define an L-structure $(\mathfrak{A}, \sigma)$ by

$$A = \{\bar{c} \mid c \in \Omega\},$$

for n-ary $R \in L$, $R^{\mathfrak{A}}\bar{c}_1 \ldots \bar{c}_n$ iff $Rc_1 \ldots c_n \in \Omega$

for n-ary $f \in L$, $f^{\mathfrak{A}}(\bar{c}_1, \ldots, \bar{c}_n) = \bar{c}$ iff $fc_1 \ldots c_n = c \in \Omega$

$$\sigma = \{\bar{U} \mid U \in \mathfrak{U}\} \text{ where } \bar{U} = \{\bar{c} \mid \text{"}c \in U\text{"} \in \Omega\}.$$

Show: (a) For atomic φ of the form $Rc_1 \ldots c_n$, $fc_1 \ldots c_n = c$, $c_1 = c_2$ or $c \in U$, one has: $(\mathfrak{A}, \sigma) \models \varphi$ iff $\varphi \in \Omega$.
(when interpreting c by $\bar{c}$ and U by $\bar{U}$).

(b) $(\mathfrak{A}, \sigma) \models \Omega$.

$(\mathfrak{A}, \sigma)$ is called the <u>term model</u> of Ω.

§ 2 The Language L_t

An L_2-formula is said to be in <u>negation normal form</u>, if negation signs in it occur only in front of atomic formulas. Using the logical rules for the negation we can assign canonically to any formula φ its negation normal form, a formula in negation normal form equivalent to φ.

An L_2-formula <u>φ is positive (negative) in the set variable X</u> if each free occurence of X in φ is within the scope of an even (odd) number of negation symbols. Equivalently, φ is positive (negative) in X, if each free occurence of X in the negation normal form of φ is of the form $t \in X$ where $t \in X$ is not preceded by a negation symbol (is of the form $\neg t \in X$). Note that for any X, which is not a free variable of φ, φ is both, positive and negative in X.

The formula

$$\exists X \neg t \in X \vee (c \in X \wedge \neg c \in Y \wedge \exists y(y \in X \wedge y \in Y))$$

is positive in X and neither positive nor negative in Y.

We use $\varphi(x_1,\ldots,x_n,X_1,\ldots,X_r)$ to denote a formula φ whose free variables are among the distinct variables $x_1,\ldots,x_n$ and whose free set variables are among the distinct set variables $X_1,\ldots,X_r$. - A simple induction shows

2.1 <u>Lemma</u>. Let $\varphi(x_1,\ldots,x_n,X_1,\ldots,X_r,Y)$ be an L_2-formula, $(\mathfrak{A},\sigma)$ a weak structure, $a_1,\ldots,a_n \in A$ and $U_1,\ldots,U_r,U \subset A$.
Assume $(\mathfrak{A},\sigma) \models \varphi[a_1,\ldots,a_n,U_1,\ldots,U_r,U]$.

(a) If φ is positive in Y, then $(\mathfrak{A},\sigma) \models \varphi[a_1,\ldots,a_n,U_1,\ldots,U_r,V]$ for any V such that $U \subset V \subset A$.

(b) If φ is negative in Y, then $(\mathfrak{A},\sigma) \models \varphi[a_1,\ldots,a_n,U_1,\ldots,U_r,V]$ for any V such that $V \subset U$.

In the sequel we use for sequences like $a_1,\ldots,a_n$ or $U_1,\ldots,U_r$ the abbreviations $\bar{a},\bar{U}$.

2.2 <u>Definition</u>. We denote by L_t the set of L_2-formulas obtained from the atomic formulas of L_2 by the formation rules of $L_{\omega\omega}$ and the rules:

(i) If t is a term and φ is positive in X, then $\forall X(t \in X \to \varphi)$ is a formula.

(ii) If t is a term and φ is negative in X, then $\exists X(t \in X \wedge \varphi)$ is a formula.

We abbreviate $\forall X(t \in X \rightarrow \varphi)$ and $\exists X(t \in X \wedge \varphi)$ by $\forall X \ni t\ \varphi$ resp. $\exists X \ni t\ \varphi$. For example,

$$\underline{\text{bas}} = \forall x\ \exists X \ni x\ \forall x\ \forall X \ni x\ \forall Y \ni x\ \exists Z \ni x\ \forall z(z \in Z \rightarrow (z \in X \wedge z \in Y))$$

is an L_t-sentence.

Note that if X is free in a subformula φ of an L_t-sentence then either φ is positive or negative in X. For an L_t-formula φ the notation $\varphi(x_1,\ldots,x_n,X_1^+,\ldots,X_r^+,Y_1^-,\ldots,Y_s^-)$ expresses that φ is positive in $X_1,\ldots,X_r$ and negative in $Y_1,\ldots,Y_s$.

2.3 __Theorem.__ L_t-sentences are invariant.

__Proof.__ For given $(\mathfrak{A},\sigma)$ one shows by induction on φ:

$$\text{if } \varphi(\bar{x},\bar{X}^+,\bar{Y}^-) \in L_t,\ \bar{a} \in A,\ \bar{U},\bar{V} \subset A, \text{ then}$$

$$(\mathfrak{A},\sigma) \models \varphi[\bar{a},\bar{U},\bar{V}] \quad \text{iff} \quad (\mathfrak{A},\tilde{\sigma}) \models \varphi[\bar{a},\bar{U},\bar{V}]\ .$$

We only treat the case $\varphi = \exists X \ni t\ \psi$. Set $a_o = t^{\mathfrak{A}}[\bar{a}]$.

Assume $(\mathfrak{A},\sigma) \models \varphi[\bar{a},\bar{U},\bar{V}]$. Choose $V \in \sigma$ such that $a_o \in V$ and $(\mathfrak{A},\sigma) \models \psi[\bar{a},\bar{U},\bar{V},V]$ By induction hypothesis, $(\mathfrak{A},\tilde{\sigma}) \models \psi[\bar{a},\bar{U},\bar{V},V]$. Hence, $(\mathfrak{A},\tilde{\sigma}) \models \varphi[\bar{a},\bar{U},\bar{V}]$. - Now suppose $(\mathfrak{A},\tilde{\sigma}) \models \varphi[\bar{a},\bar{U},\bar{V}]$. Let $V \in \tilde{\sigma}$ be such that $a_o \in V$ and $(\mathfrak{A},\tilde{\sigma}) \models \psi[\bar{a},\bar{U},\bar{V},V]$. By induction hypothesis, $(\mathfrak{A},\sigma) \models \psi[\bar{a},\bar{U},\bar{V},V]$: Since $V \in \tilde{\sigma}$, there is a $V' \in \sigma$ such that $a_o \in V' \subset V$. ψ is negative in X because $\exists X \ni t\ \psi \in L_t$. Thus by 2.1, $(\mathfrak{A},\sigma) \models \psi[\bar{a},\bar{U},\bar{V},V']$, hence $(\mathfrak{A},\sigma) \models \varphi[\bar{a},\bar{U},\bar{V}]$.

2.4 __Corollary.__ Suppose that σ_1 and σ_2 are bases of the same topology on $A, \tilde{\sigma}_1 = \tilde{\sigma}_2$. Let φ be an L_t-sentence. Then

$$(\mathfrak{A},\sigma_1) \models \varphi \quad \text{iff} \quad (\mathfrak{A},\sigma_2) \models \varphi\ .$$

The properties "Hausdorff", "regular", "discrete" and "trivial" of topologies may be expressed by L_t-sentences (though the sentences φ_{haus} and φ_{disc} of the last section are not in L_t):

$$\underline{\text{haus}} = \forall x\ \forall y\ (x = y \vee \exists X \ni x\ \exists Y \ni y\ \forall z \neg (z \in X \wedge z \in Y))$$
$$\underline{\text{reg}} = \forall x\ \forall X \ni x\ \exists Y \ni x\ \forall y\ (y \in X \vee \exists W \ni y\ \forall z\ (\neg z \in W \vee \neg z \in Y))$$
$$\underline{\text{disc}} = \forall x\ \exists X \ni x\ \forall y\ (y \in X \rightarrow y = x)$$
$$\underline{\text{triv}} = \forall x\ \forall X \ni x\ \forall y\ y \in X\ .$$

For an n-ary function symbol $f \in L$ the continuity of f is expressed in L_t by

$$\varphi = \forall x_1 \ldots \forall x_n \ \forall Y \ni fx_1 \ldots x_n \ \exists X_1 \ni x_1 \ldots \exists X_n \ni x_n$$

$$\forall y_1 \ldots \forall y_n (y_1 \in X_1 \wedge \ldots \wedge y_n \in X_n \rightarrow fy_1 \ldots y_n \in Y),$$

i.e. one has for all topological structures $(\mathfrak{A},\sigma)$

$(\mathfrak{A},\sigma) \models \varphi$ iff f^A is a continuous map from A^n to A (where A^n carries the product topology).

The class of topological groups and the class of topological fields are axiomatizable in L_t; for example, if $L = \{\cdot, ^{-1}, e\}$ then the topological groups are just the structures which are models of the group axioms and the sentences " $\cdot$ is continuous", and " $^{-1}$ is continuous".

By <u>topological model theory</u> (or topological logic) we understand the study of topological structures using the formal language L_t (and variants of L_t).

2.5 <u>Exercise</u>. (a) Show that for unary $f \in L$, "f is an open map" may be expressed in L_t (compare 1.3 (b)).
(b) Show that for unary $P \in L$, "P is open" may be expressed in L_t (compare 1.3 (c)).
(c) Show that for $\varphi \in L_t$ there is a $\psi \in L_{\omega\omega}$ such that for all topological structures $(\mathfrak{A},\sigma)$ with $(\mathfrak{A},\sigma) \models$ <u>disc</u> one has:

$$(\mathfrak{A},\sigma) \models \varphi \quad \text{iff} \quad \mathfrak{A} \models \psi .$$

Similarly for models of <u>triv</u>.

<u>§ 3 Beginning topological model theory</u>

Using the invariance of the sentences of L_t one can derive many theorems for topological logic from its classical analogues. This section contains some examples.

Given $\Phi \cup \{\varphi\} \subset L_2$ we write $\Phi \models \varphi$ resp. $\Phi \models_t \varphi$ if each weak structure resp. topological structure that is a model of Φ is a model of φ.

3.1 <u>Lemma</u>. Suppose $\Phi \cup \{\varphi\} \subset L_t$.
(a) Φ has a topological model iff $\Phi \cup \{$<u>bas</u>$\}$ has a weak model.

(b) $\Phi \models_t \varphi$ iff $\Phi \cup \{$<u>bas</u>$\} \models \varphi$.

<u>Proof</u>. (a): If Φ has a topological model $(\mathfrak{A},\sigma)$, then $(\mathfrak{A},\sigma) \models \Phi \cup \{\underline{bas}\}$. - Conversely, suppose that the weak structure $(\mathfrak{A},\sigma)$ is a model of $\Phi \cup \{\underline{bas}\}$. Since $(\mathfrak{A},\sigma) \models \underline{bas}$, $\tilde{\sigma}$ is a topology on A. Since $(\mathfrak{A},\sigma) \models \Phi$ we get, by invariance of L_t-sentences, $(\mathfrak{A},\tilde{\sigma}) \models \Phi$. - (b) is easily derived from (a).

Using 3.1 we obtain

3.2 <u>Compactness theorem</u>. A set of L_t-sentences has a topological model if every finite subset does.

3.3 <u>Completeness theorem</u>. For recursive L, the set of L_t-sentences which hold in all topological structures is recursively enumerable.

We say that a topological structure $(\mathfrak{A},\sigma)$ is <u>denumerable</u>, if A is denumerable (i.e. finite or countable) and σ has a denumerable basis.

3.4 <u>Löwenheim-Skolem theorem</u>. A denumerable set Φ of L_t-sentences which has a topological model has a denumerable topological model.

<u>Proof</u>. By assumption and 3.1 (a), $\Phi \cup \{\underline{bas}\}$ is satisfiable. Thus, by Löwenheim-Skolen theorem for L_2, there is a weak model of $(\mathfrak{A},\sigma)$ such that $A \cup \sigma$ is denumerable. Then, $(\mathfrak{A},\tilde{\sigma})$ is a denumerable topological model of Φ.

3.5 <u>Corollary</u>. The class of normal spaces cannot be axiomatized in L_t.

<u>Proof</u>. Suppose $\varphi_o \in L_t$ axiomatizes the class of normal spaces. Let (B,τ) be a regular but not normal space, i.e. $(B,\tau) \models \underline{reg} \wedge \neg \varphi_o$. By 3.4 there is a denumerable topological model (A,σ) of $\underline{reg} \wedge \neg \varphi_o$. Since (A,σ) is denumerable and regular it is metrizable, hence normal, which contradicts $(\mathfrak{A},\sigma) \models \neg \varphi_o$.

3.6 <u>Corollary</u>. The class of connected spaces cannot be axiomatized in L_t.

<u>Proof</u>. Each connected and ordered topological field is isomorphic to the field of real numbers, and hence is uncountable.

3.7 <u>Exercise</u>. Show that the class of compact spaces cannot be axiomatized in L_t.

We do not state the Löwenheim-Skolem-Tarski theorem for topological logic but we use it in the following

3.8 <u>Exercise</u>. Suppose (A,τ) is a T_3-space (i.e. Hausdorff and regular) with countable A. Show: If σ_o is a countable subset of τ, then there is a T_3-

topology σ such that $\sigma_o \subset \sigma \subset \tau$ and σ has a countable basis.

Similarly, show that a space with a countable basis is regular iff each countable subspace is regular.

A set of L_t-sentences is called an L_t-theory. We denote theories by $T, T', \ldots$. - Using 3.1 one can obtain two cardinal theorems for topological logic. We only state one result:

3.9 Theorem. Let $(\mathfrak{A},\sigma)$ be a topological model of an L_t-theory T. Suppose that the cardinality $|A|$ of A is a regular cardinal κ and that each point of A has a neighborhood basis of less than κ sets. Then T has a topological model whose universe has cardinality $\aleph_1$, and such that each point has a denumerable neighborhood basis.

Proof. Let $<^A$ be a well-ordering of A of type κ. Choose $\sigma' \subset \sigma$ such that σ' contains, for each $a \in A$, a basis of neighborhoods of a cardinality less than κ. Take a new ternary relation symbol R and choose an interpretation R^A of R in A such that

$$(\mathfrak{A},<^A,R^A,\sigma') \models \varphi_o ,$$

where $\varphi_o = \forall x \forall z \exists X \ni x \forall u(u \in X \leftrightarrow Rxzu) \wedge$
$$\forall x \exists y \forall X(x \in X \leftrightarrow \exists z(z < y \wedge Rxzx \wedge \forall u(Rxzu \rightarrow u \in X)))$$

(i.e. $\{Rxz\cdot \mid z < y\}$ is a basis of x).

By a classical two-cardinal theorem there is a model $(\mathfrak{B},<^B,R^B,\tau)$ of $T \cup \{\underline{bas}\} \cup \{\varphi_o\}$ with $|B| = \aleph_1$, and $\aleph_1$-like ordering $<^B$. Since $(\mathfrak{B},<^B,R^B,\tau) \models \underline{bas} \wedge \varphi_o$, τ is the basis of a topology in which each point has a denumerable neighborhood basis, i.e. $(\mathfrak{B},\tilde{\tau})$ is the desired model of T.

3.10 Exercise. Suppose that the denumerable L_t-theory T has an infinite topological model. Show that T has a denumerable topological model with $2^{\aleph_o}$ homeomorphisms.

The infinitary languages $(L_{\infty\omega})_t$ and $(L_{\omega_1\omega})_t$ are obtained from L_t by adding the formation rule:

If Φ is an arbitrary resp. countable set of formulas, then $\bigwedge\Phi$ and $\bigvee\Phi$ are formulas (the conjunction and disjunction of the formuals in Φ).

$(L_{\infty\omega})_t$-sentences are invariant; hence the analogue of 3.1 holds for

$\Phi \subset (L_{\infty\omega})_t$,

Φ has a topological model iff $\Phi \cup \{\underline{bas}\}$ has a weak model.

Using this fact, we can generalize classical results; for example, we get the Löwenheim-Skolem theorem for $(L_{\omega_1\omega})_t$; we can show that the class of well-orderings is not axiomatizable by an $(L_{\infty\omega})_t$-sentence, and that the well-ordering number of $(L_{\omega_1\omega})_t$ is $\aleph_1$.

Much more involved and sometimes even impossible are the proofs for L_t, $(L_{\omega_1\omega})_t$ and $(L_{\infty\omega})_t$ of theorems which - as the interpolation theorem, the omitting types theorem or Scotts isomorphism theorem - claim the existence of a formula having a certain property.

3.11 Exercise. Let $L_t(Q)$ be the language obtained from L_t by adding the quantifier Qx expressing "there are uncountable many x". Show, using the corresponding results for $L_{\omega\omega}(Q)$, that if we restrict to topological structures $L_t(Q)$ satisfies the compactness theorem for countable sets of sentences and the completeness theorem.

3.12 Exercise. Show that for $L = \emptyset$ there is no set T of L_t-sentences such that the class of topological models of T are just the topological spaces carrying a topology induced by a uniformity.

Appendix: Many-sorted languages

Sometimes the exposition will be easier if we use many-sorted weak structures, i.e. structures of the form

$$((\mathfrak{A}^1,\sigma^1),\ldots,(\mathfrak{A}^r,\sigma^r),\ldots).$$

Here we sketch the definitions for the notions we need, when discussing such structures:

Let S be a non-empty finite set, $S = \{i_1,\ldots,i_r\}$, the set of sorts, and let L^o be a similarity type. Assume that typ is a function associating sorts with each symbol in L^o: if $R \in L^o$ is n-ary, then typ(R) is an n-tuple $(j_1,\ldots,j_n) \in S^n$; and if $f \in L^o$ is n-ary, then typ(f) is an (n+1)-tuple $(j_1,\ldots,j_n,j) \in S^{n+1}$. $L = (L^o,S,\text{typ})$ is called a many sorted-similarity type.

For $i \in S$ we denote by L(i) the one-sorted similarity type

$$L(i) = \{k \mid k \in L^o, \text{typ}(k) = (i,\ldots,i)\}.$$

Now, a many-sorted weak L-structure $\mathfrak{A}$ consists of weak L(i)-structures $(\mathfrak{A}^i,\tau^i)$ for $i \in S$, and of the interpretations $k^{\mathfrak{A}}$ of symbols $k \in L^o - \bigcup_{i\in S} L(i)$: if $k = R$, and $typ(R) = (j_1,\dots,j_n)$ then $R^{\mathfrak{A}} \subset A^{j_1}\times\dots\times A^{j_n}$, and if $k = f$, and $typ(f) = (j_1,\dots,j_n,j)$ then $f^{\mathfrak{A}}: A^{j_1}\times\dots\times A^{j_n} \to A^j$. We write $\mathfrak{A}$ also in the form

$$((\mathfrak{A}^{i_1},\tau^{i_1}),\dots,(\mathfrak{A}^{i_r},\tau^{i_r}),(k^{\mathfrak{A}})_{k \in L^o - \bigcup_{i\in S} L(i)}) .$$

We call $\mathfrak{A}$ a <u>topological structure</u> if all $(\mathfrak{A}^i,\tau^i)$ are topological structures.

Now we define for a many-sorted similarity type $L = (L^o,S,typ)$ the languages L_2 and L_t: For each sort i we introduce countably many variables $(x^i,y^i,z^i,\dots)$ and countably many set variables $(X^i,Y^i,Z^i,\dots)$. The terms of sort i are the variables $x^i,y^i,\dots$, and expressions of the form $ft_1,\dots t_n$, where $typ(f) = (j_1,\dots,j_n,i)$, and each t_s is of sort j_s. The atomic formulas are

$t_1 = t_2$	if t_1,t_2 are terms of the same sort.
$Rt_1\dots t_n$	where $typ(R) = (j_1,\dots,j_n)$ and t_s is a term of sort j_s.
$t \in X^i$	where t is a term of sort i.

The formulas are obtained closing under the logical operations $\neg,\wedge,\vee,\exists x^i,\forall x^i$, $\exists X^i$ and $\forall X^i$, where in case of L_t we have the following restrictions for $\exists X^i$ and $\forall X^i$:

If t is a term of sort i and φ is positive (negative) in X^i, then $\forall X^i(t \in X^i \to \varphi)$ $(\exists X^i(t \in X^i \wedge \varphi))$ is a formula.

L_t-formulas are invariant (where the concept of invariance is defined in the obvious way).

For example, the class of topological vector spaces is axiomatizable by L_t-sentences using a two-sorted L, one sort for the scalars (i.e. elements of a topological field) and one sort for the vectors. - Show that the class of sheaves is L_t-axiomatizable in an appropriate many-sorted L.

Sometimes in the next sections, we introduce for one-sorted L-structures $(\mathfrak{A},\sigma)$ and $(\mathfrak{B},\tau)$,a many-sorted structure of type $((\mathfrak{A},\sigma),(\mathfrak{B},\tau),\dots)$. We are going to use this notation, though to be precise we should introduce a copy L* of L disjoint from L and look at the structure $((\mathfrak{A},\sigma),(\mathfrak{B}^*,\tau),\dots)$ where $\mathfrak{B}^*$ is the L*-structure corresponding to $\mathfrak{B}$.
From now on, unless otherwise stated, all similarity types will be one-sorted.

3.13 Exercise. Show that there is no L_t-sentence which expresses " f is a closed map" (for, say, unary f) in all topological structures (Hint: Look at the structure $\mathfrak{A} = ((\mathbb{N} \times \mathbb{N}, \sigma), (\mathbb{N}, <, \tau), p^A, f^A)$, where p is binary and f is unary.

$p^A : \mathbb{N} \times \mathbb{N} \to \mathbb{N} \times \mathbb{N}$ is defined by $p^A(x,y) = (x,y)$

$f^A : \mathbb{N} \times \mathbb{N} \to \mathbb{N}$ is defined by $f^A((x,y)) = y$.

$\tau = \{U | U \subset (\mathbb{N} - \{0\}) \text{ or } \mathbb{N} - U \text{ is finite}\}$,

and for $U \subset \mathbb{N} \times \mathbb{N}$,

$U \in \sigma$ iff U satisfies (i) and (ii),

where (i) if $(n,m) \in U$ then for some r, $\{(n,s) | s \geq r\} \subset U$,

(ii) if $(n,0) \in U$ then for some r, $\{(s,l) | s \geq r, l \in \mathbb{N}\} \subset U$.

f^A is a closed map, but $f^{\mathfrak{B}}$ is not closed for any countable "non-standard" topological structure $\mathfrak{B}$ satisfying the same L_t-sentence as $\mathfrak{A}$.

§ 4 Ehrenfeucht-Fraïssé Theorem

We call $(\mathfrak{A},\sigma)$ and $(\mathfrak{B},\tau)$ L_t-equivalent, in symbols $(\mathfrak{A},\sigma) \equiv^t (\mathfrak{B},\tau)$, if they satisfy the same L_t-sentences. This section contains an algebraic characterization of L_t-equivalence, the analogue of the classical Ehrenfeucht-Fraïssé theorem. For this we introduce back and forth methods for L_t which also will be of use later on.

Extending the terminology of topology, we call weak structures $(\mathfrak{A},\sigma)$ and $(\mathfrak{B},\tau)$ homeomorphic, written $(\mathfrak{A},\sigma) \simeq^t (\mathfrak{B},\tau)$, if $(\mathfrak{A},\tilde{\sigma})$ and $(\mathfrak{B},\tilde{\tau})$ are isomorphic.

Suppose π^0 is an isomorphism of $(\mathfrak{A},\tilde{\sigma})$ onto $(\mathfrak{B},\tilde{\tau})$. If we define the binary relations π^1 and $\pi^2, \pi^1, \pi^2 \subset \sigma \times \tau$, by

$$\pi^1 = \{(U,V) | (U,V) \in \sigma \times \tau, \pi^0(U) \subset V\}$$

$$\pi^2 = \{(U,V) | (U,V) \in \sigma \times \tau, (\pi^0)^{-1}(V) \subset U\},$$

then the following holds:

(1) π^0 is an isomorphism of $\mathfrak{A}$ onto $\mathfrak{B}$.

(2) If $U \pi^1 V$ and $a \in U$ then $\pi^o(a) \in V$.
If $U \pi^2 V$, $\pi^o(a) = b$ and $b \in V$ then $a \in U$.
(We write $U \pi^i V$ for all $(U,V) \in \pi^i$).

(3) For each $a \in A$ and $V \in \tau$ with $\pi^o(a) \in V$ there is a $U \in \sigma$ such that $a \in U$ and $U \pi^1 V$.

(4) For $a \in A$ and $U \in \sigma$ with $a \in U$ there is a $V \in \tau$ such that $\pi^o(a) \in V$ and $U \pi^2 V$.

Conversely, if for $(\mathfrak{A},\sigma)$ and $(\mathfrak{B},\tau)$ there are a map π^o and relations $\pi^1,\pi^2 \subset \sigma \times \tau$ satisfying (1) - (4), then π^o is an isomorphsm of $(\mathfrak{A},\tilde{\sigma})$ onto $(\mathfrak{B},\tilde{\tau})$. We then write $(\pi^o,\pi^1,\pi^2): (\mathfrak{A},\sigma) \simeq^t (\mathfrak{B},\tau)$.

Note that an L_2-sentence φ is invariant if and only if it is preserved under $\simeq^t$, the relation of homeomorphism, i.e.

$$\text{if } (\mathfrak{A},\sigma) \models \varphi \text{ and } (\mathfrak{A},\sigma) \simeq^t (\mathfrak{B},\tau) \text{ then } (\mathfrak{B},\tau) \models \varphi.$$

We are going to define a countable approximation of $\simeq^t$ (the relation of partial homeomorphism) and finite approximations of $\simeq^t$. (Compare [6] where for the relations of isomorphism, homomorphism ... of classical model theory such approximations were defined and used for model-theoretic purposes).The finite approximations will be definable by L_t-sentences.From this we obtain the analogue of the Ehrenfeucht-Fraïssé theorem. By the way, we get another algebraic characterization of L_t-equivalence: two topological structures are L_t-equivalent iff they have homeomorphic ultrapowers. We close the section with a further application of back and forth methods showing that the L_t-sentences are the invariant L_2-sentences.

4.1 <u>Definition</u>. $p = (p^o,p^1,p^2)$ is a <u>partial homeomorphism</u> from $(\mathfrak{A},\sigma)$ to $(\mathfrak{B},\tau)$, in symbols $p \in P((\mathfrak{A},\sigma),(\mathfrak{B},\tau))$, if

(i) p^o is a partial isomorphism from $\mathfrak{A}$ to $\mathfrak{B}$, i.e. a one-to-one mapping with $dom(p^o) \subset A$, $rg(p^o) \subset B$ such that for $R \in L$, $\bar{a} \in dom(p^o)$

$$R^{\mathfrak{A}}\bar{a} \quad \text{iff} \quad R^{\mathfrak{B}}p^o(\bar{a})$$

($p^o(\bar{a})$ denotes $p^o(a_o),\ldots,p^o(a_{n-1})$ if $\bar{a}$ is $a_o\ldots a_{n-1}$),

and for $f \in L$, $\bar{a}$, $a \in dom(p^o)$

$$f^{\mathfrak{A}}(\bar{a}) = a \quad \text{iff} \quad f^{\mathfrak{B}}(p^o(\bar{a})) = p^o(a).$$

(ii) p^1 and p^2 are relations, $p^1, p^2 \subset \sigma \times \tau$, satisfying:

if $U\, p^1\, V$, $a \in dom(p^o)$ and $a \in U$ then $p^o(a) \in V$,

if $U\, p^2\, V$, $b \in rg(p^o)$, say $p^o(a) = b$, and $b \in V$ then $a \in U$.

Given $p = (p^o, p^1, p^2)$, $q = (q^o, q^1, q^2) \in P((\mathfrak{A},\sigma),(\mathfrak{B},\tau))$ we write $p \subset q$, if $p^i \subset q^i$ holds for $i = 0,1,2$.

4.2 Definition. We write $I: (\mathfrak{A},\sigma) \simeq_p^t (\mathfrak{B},\tau)$ and say that $(\mathfrak{A},\sigma)$ and $(\mathfrak{B},\tau)$ are partially homeomorphic via I, if I is a non-empty set of partial homeomorphisms with the following back and forth properties:

(forth_1) For $p \in I$ and $a \in A$ there is $q \in I$ with $p \subset q$ and $a \in dom(q^o)$.

(forth_2) For $p \in I$, $a \in dom(p^o)$ and $U \in \sigma$ with $a \in U$ there are $q \in I$ and $V \in \tau$ such that $p \subset q$, $p^o(a) \in V$ and $U\, q^2\, V$.

(back_1) For $p \in I$ and $b \in B$ there is $q \in I$ with $p \subset q$ and $b \in rg(q^o)$.

(back_2) For $p \in I$, $b \in rg(p^o)$, say $p^o(a) = b$, and $V \in \tau$ with $b \in V$ there are $q \in I$ and $U \in \sigma$ such that $p \subset q$, $a \in U$ and $U\, q^1\, V$.

We write $(\mathfrak{A},\sigma) \simeq_p^t (\mathfrak{B},\tau)$ if there is an I such that $I: (\mathfrak{A},\sigma) \simeq_p^t (\mathfrak{B},\tau)$.

The relation $\simeq_p^t$ is a countable approximation of $\simeq^t$ in the following sense.

4.3 Lemma. (a) If $(\mathfrak{A},\sigma) \simeq^t (\mathfrak{B},\tau)$ then $(\mathfrak{A},\sigma) \simeq_p^t (\mathfrak{B},\tau)$.

(b) If $(\mathfrak{A},\sigma) \simeq_p^t (\mathfrak{B},\tau)$ and $A \cup \sigma \cup B \cup \tau$ is denumerable, then $(\mathfrak{A},\sigma) \simeq^t (\mathfrak{B},\tau)$.

Proof. (a): Suppose $(\pi^o, \pi^1, \pi^2) : (\mathfrak{A},\sigma) \simeq^t (\mathfrak{B},\tau)$. Then $I: (\mathfrak{A},\sigma) \simeq_p^t (\mathfrak{B},\tau)$ holds for $I = \{(\pi^o, \pi^1, \pi^2)\}$.

(b): Assume $I: (\mathfrak{A},\sigma) \simeq_p^t (\mathfrak{B},\tau)$ where $A \times \sigma = \{(a_i, U_i) \mid i \in \omega\}$ and $B \times \tau = \{(b_i, V_i) \mid i \in \omega\}$. We define $p_n \in I$ by induction. Let p_o be an arbitrary element of I. Given p_n choose, using the back and forth properties, $p_{n+1} \in I$ such that

$p_n \subset p_{n+1}$

$a_n \in dom(p^o_{n+1})$, $b_n \in rg(p^o_{n+1})$

if $a_n \in U_n$ then there is a $V \in \tau$ such that $p^o_{n+1}(a_n) \in V$

and $U_n p^2_{n+1} V$.

if $b_n \in V_n$, and $p^o_{n+1}(a) = b_n$ then there is a $U \in \sigma$ such that $a \in U$ and $U \; p^1_{n+1} \; V_n$.

For $i = 0,1,2$ set $p^i = \bigcup_n p^i_n$. Then $(p^o,p^1,p^2): (\mathfrak{A},\sigma) \simeq^t (\mathfrak{B},\tau)$.

4.4 Remarks. 1) (Ehrenfeucht games). Given structures $(\mathfrak{A},\sigma)$ and $(\mathfrak{B},\tau)$ we introduce an infinite two person game $G((\mathfrak{A},\sigma),(\mathfrak{B},\tau))$ with players I and II. There are countably many moves and each move is of type x or of type X. In the n-th move, player I first decides which type this move is to be. In an x-move player I picks $a_n \in A$ (or $b_n \in B$), and then II picks $b_n \in B$ (resp. $a_n \in A$). In an X-move player I chooses a $U_n \in \sigma$ and an $i < n$ such that the i-th move was of type x and $a_i \in U_n$ (or $V_n \in \tau$ with $b_i \in V_n$); player II then has to choose $V_n \in \tau$ with $b_i \in V_n$ (resp. $U_n \in \sigma$ with $a_i \in U_n$). II wins if (p^o,p^1,p^2) is a partial homeomorphism where

$p^o = \{(a_n,b_n) \mid n \in \omega$, the n-th move was of type x and the elements a_n and b_n were chosen$\}$,

$p^1 = \{(U_n,V_n) \mid n \in \omega$, the n-th move was of type X, I chose V_n and II chose $U_n\}$,

$p^2 = \{(U_n,V_n) \mid n \in \omega$, the n-th move was of type X, I chose U_n and II chose $V_n\}$.

Now, it is easy to prove that

$$(\mathfrak{A},\sigma) \simeq^t_p (\mathfrak{B},\tau) \quad \text{iff} \quad \text{II has a winning strategy in the game } G((\mathfrak{A},\sigma),(\mathfrak{B},\tau)).$$

2) For any weak structures $(\mathfrak{A},\sigma),(\mathfrak{B},\tau)$, we have

$$(\mathfrak{A},\sigma) \simeq^t_p (\mathfrak{B},\tau) \quad \text{iff} \quad (\mathfrak{A},\tilde{\sigma}) \simeq^t_p (\mathfrak{B},\tilde{\tau}).$$

Proof. For $p \in P((\mathfrak{A},\sigma),(\mathfrak{B},\tau))$ put

$$\hat{p} = (p^o, \{(U',V') \mid (U',V') \in \tilde{\sigma} \times \tilde{\tau}, \text{there is } (U,V) \in p^1 \text{ such that } U' \subset U, V \subset V'\},$$
$$\{(U',V') \mid (U',V') \in \tilde{\sigma} \times \tilde{\tau}, \text{ there is } (U,V) \in p^2 \text{ such that } U \subset U', V' \subset V\}),$$

and for a set I let $\hat{I} = \{\hat{p} \mid p \in I\}$.

Now, if $I: (\mathfrak{A},\sigma) \simeq^t_p (\mathfrak{B},\tau)$ then $\hat{I}: (\mathfrak{A},\tilde{\sigma}) \simeq^t_p (\mathfrak{B},\tilde{\tau})$.

Conversely, suppose $I: (\mathfrak{A},\tilde{\sigma}) \simeq^t_p (\mathfrak{B},\tilde{\tau})$. Then, in particular, $\hat{I}: (\mathfrak{A},\tilde{\sigma}) \simeq^t_p (\mathfrak{B},\tilde{\tau})$. One easily proves that $\hat{I} \upharpoonright \sigma \times \tau: (\mathfrak{A},\sigma) \simeq^t_p (\mathfrak{B},\tau)$ where

$$\hat{I} \upharpoonright \sigma \times \tau = \{(p^o, p^1 \cap (\sigma \times \tau), p^2 \cap (\sigma \times \tau)) \mid (p^o,p^1,p^2) \in \hat{I}\}.$$

3) Show: $(\mathfrak{A},\sigma) \simeq^t_p (\mathfrak{B},\tau)$ holds just in case there is a non-empty set I of par-

tial homoeomorphisms with the properties (forth$_1$),(back$_1$),(forth*) and (back*) where

(forth*) For $p \in I$, $a \in dom(p^o)$ and $U \in \sigma$ with $a \in U$ there are $q \in I$, $U' \in \sigma$, $V \in \tau$ such that $p \subset q$, $a \in U' \subset U$, $p^o(a) \in V$ and $U' \, q^2 \, V$.

(back*) For $p \in I$, $b \in rg(p^o)$, say $p^o(a) = b$, and $V \in \tau$ with $b \in V$ there are $q \in I$, $U \in \sigma$, $V' \in \tau$ such that $p \subset q$, $a \in U$, $b \in V' \subset V$ and $U \, q^1 \, V'$.

4) Given $(\mathfrak{A},\sigma)$ and $(\mathfrak{B},\tau)$ with $(\mathfrak{A},\sigma) \simeq_p^t (\mathfrak{B},\tau)$, $A \in \sigma$ and $B \in \tau$ there is an I such that $I: (\mathfrak{A},\sigma) \simeq_p^t (\mathfrak{B},\tau)$ and $(A,B) \in p_1$. $(A,B) \in p^2$ for all $p \in I$. The same is true for $(\emptyset,\emptyset)$, if $\emptyset \in \sigma$ and $\emptyset \in \tau$.

4.5 Exercise. Let L be one-sorted. Show that there is a many-sorted L' and a set Φ_p of L_t'-sentences such that for any L-structures $(\mathfrak{A}_1,\sigma_1)$ and $(\mathfrak{A}_2,\sigma_2)$, we have:

$(\mathfrak{A}_1,\sigma_1) \simeq_p^t (\mathfrak{A}_2,\sigma_2)$ iff $((\mathfrak{A}_1,\sigma_1),(\mathfrak{A}_2,\sigma_2),\ldots) \models \Phi_p$ for some choice of the universes and relations in ... .

(Hint: Take $S = \{1,2,3,4,5\}$ as set of sorts. Look at the class of structures of the form

$$((\mathfrak{A}_1,\sigma_1),(\mathfrak{A}_2,\sigma_2),(F_1,\sigma_3),(F_2,\sigma_4),(I,\sigma_5),E_1,E_2, ---),$$

which are models of the L_t'-sentences (note that σ_3,σ_4 and σ_5 will be arbitrary):

"F_1 is via E_1 a basis of $\tilde{\sigma}_1$"

(i.e. $\forall x^1 \, \forall X^1 \ni x^1 \, \exists x^3 (E_1 x^1 x^3 \wedge \forall y^1 (E y^1 x^3 \rightarrow y^1 \in X^1))$
$\wedge \forall x^1 \, \forall x^3 (E_1 x^1 x^3 \rightarrow \exists X^1 \ni x^1 \, \forall y^1 (y^1 \in X^1 \rightarrow E_1 y^1 x^3)))$.

"F_2 is via E_2 a basis of $\tilde{\sigma}_2$"

" I is via the relations in --- a set of partial homeomorphisms with $I: (\mathfrak{A}_1,F_1) \simeq_p^t (\mathfrak{A}_2,F_2)$".

Note that introducing F_1 and F_2 we are able to quantify in L_t' arbitrarily over the elements of a basis of $\tilde{\sigma}$ resp. $\tilde{\tau}$; in particular, we can formulate the back and forth properties. Note also that by the preceding, every PC-class over L_2 definable by an invariant sentence, is a PC-class over L_t.

4.6 Lemma. Partially homeomorphic structures are L_t-equivalent.

Proof. Assume I: $(\mathfrak{A},\sigma) \simeq_p^t (\mathfrak{B},\tau)$. By induction on $\psi \in L_t$ we show

if ψ is in negation normal form $\psi = \psi(x_o,\ldots,x_{n-1},X_o^+,\ldots,X_{r-1}^+,$
$Y_o^-,\ldots,Y_{s-1}^-)$, and if $p \in I$, $\{(a_i,b_i)| i < n\} \subset p^o$,
$\{(U_i^+,V_i^+)| i < r\} \subset p^1$ and $\{(U_i^-,V_i^-)| i < s\} \subset p^2$, then
$(\mathfrak{A},\sigma) \models \psi[\bar{a},\bar{U}^+,\bar{U}^-]$ implies $(\mathfrak{B},\tau) \models \psi[\bar{b},\bar{V}^+,\bar{V}^-]$

Two examples should suffice. Let ψ be of the form $x_i \in Z$. Then Z occurs positively in ψ, i.e. $Z = X_j$ for some $j < r$. If $(\mathfrak{A},\sigma) \models x_i \in X_j[\bar{a},\bar{U}^+,\bar{U}^-]$ then $a_i \in U_j^+$. Since $p^o(a_i) = b_i$ and $U_j^+ \, p^1 \, V_j^+$, we have $b_i \in V_j^+$. Hence $(\mathfrak{B},\tau) \models \psi[\bar{b},\bar{V}^+,\bar{V}^-]$.

Now, let ψ be $\exists Y \ni x_i\, \varphi$. If $(\mathfrak{A},\sigma) \models \psi[\bar{a},\bar{U}^+,\bar{U}^-]$ then there is $U \in \sigma$ with $a_i \in U$ and $(\mathfrak{A},\sigma) \models \varphi[\bar{a},\bar{U}^+,\bar{U}^-,U]$.

Using (forth$_2$) we obtain $q \in I$ and $V \in \tau$ such that $p \subset q, b_i \in V$ and $U \, q^2 \, V$. Then by induction hypothesis, $(\mathfrak{B},\tau) \models \varphi[\bar{b},\bar{V}^+,\bar{V}^-,V]$. But $b_i \in V$, hence $(\mathfrak{B},\tau) \models \psi[\bar{b},\bar{V}^+,\bar{V}^-]$.

We call a weak structure $(\mathfrak{A},\sigma)$ ω-saturated, recursively saturated,... if it is ω-saturated, recursively saturated,... as a two-sorted structure (and we do similarly in case of structures of type $((\mathfrak{A},\sigma),(\mathfrak{B},\tau))$). Clearly, we assume that L is recursive when speaking of a recursively saturated structure. Though not so convenient for our present purposes, the more natural definition in topological model theory, say of a recursively saturated topological structure $(\mathfrak{A},\sigma)$ would be: $(\mathfrak{A},\sigma)$ is recursively saturated iff for some basis τ of σ the two-sorted structure $(\mathfrak{A},\tau)$ is recursively saturated.

Though the following converse of 4.6 will be an immediate consequence of the L_t-Ehrenfeucht-Fraïssé theorem, we give here a direct proof.

4.7 Lemma. Suppose that $(\mathfrak{A},\sigma)$ and $(\mathfrak{B},\tau)$ are ω-saturated (or that $((\mathfrak{A},\sigma),(\mathfrak{B},\tau))$ is recursively saturated). Then $(\mathfrak{A},\sigma) \equiv^t (\mathfrak{B},\tau)$ implies $(\mathfrak{A},\sigma) \simeq_p^t (\mathfrak{B},\tau)$. In particular, if $A \cup B \cup \sigma \cup \tau$ is denumerable, then $(\mathfrak{A},\sigma) \simeq^t (\mathfrak{B},\tau)$.

Proof. Assume $(\mathfrak{A},\sigma) \equiv^t (\mathfrak{B},\tau)$. Let I be the set

$I = \{p| \, p$ has the form $(\{(a_i,b_i)| i < n\},\{(U_i^+,V_i^+)| < r\},\{(U_i^-,V_i^-)|i < s\})$,

and for $\psi(x_o,\ldots,x_{n-1},X_o^+,\ldots,X_{r-1}^+,Y_o^-,\ldots,Y_{s-1}^-) \in L_t$ the following holds: if $(\mathfrak{A},\sigma) \models \psi[\bar{a},\bar{U}^+,\bar{U}^-]$ then $(\mathfrak{B},\tau) \models \psi[\bar{b},\bar{V}^+,\bar{V}^-]\}$.

We proof I: $(\mathfrak{A},\sigma) \simeq_p^t (\mathfrak{B},\tau)$: Since $(\mathfrak{A},\sigma) \equiv^t (\mathfrak{B},\tau)$, $(\emptyset,\emptyset,\emptyset)$ is in I, hence I is non-empty. Choosing appropriate atomic or negated atomic $\psi \in L_t$, one easily shows that each p in I is a partial homeomorphism.

Let us check (back_2) for I in case $((\mathfrak{A},\sigma),(\mathfrak{B},\tau))$ is recursively saturated. Suppose $p \in I$, say $p = (\{a_i,b_i)|i < n\}, \{(U_i^+,V_i^+)|i < r\},\{(U_i^-,V_i^-)|i < s\})$. Let $b \in rg(p^o)$, i.e. $b = b_i$ for some i, and take any $V \in \tau$ with $b_i \in V$. We set

$$\Phi = \{\varphi|\varphi \in L_t,\varphi = \varphi(x_o,\ldots,x_{n-1},X_o^+,\ldots,X_{r-1}^+,X^+,Y_o^-,\ldots,Y_{s-1}^-)\}.$$

To each finitely many $\varphi_1,\ldots,\varphi_e \in \Phi$ with

$$(*) \qquad (\mathfrak{B},\tau) \models \neg\varphi_1\wedge\ldots\wedge\neg\varphi_e[\bar{b},\bar{V}^+,V,\bar{V}^-] \quad ,$$

there is a $U' \in \sigma$ such that $a_i \in U'$ and

$$(\mathfrak{A},\sigma) \models \neg\varphi_1\wedge\ldots\wedge\neg\varphi_e[\bar{a},\bar{U}^+,U',\bar{U}^-].$$

Otherwise, $(\mathfrak{A},\sigma) \models \forall X \ni x_i\ \varphi_1\vee\ldots\vee\varphi_e[\bar{a},U^+,U^-]$, thus $(\mathfrak{B},\tau) \models \forall X \ni x_i\ \varphi_1\vee\ldots\vee\varphi_e[\bar{b},V^+,V^-]$ since $p \in I$. But this contradicts (*).

Therefore, the following recursive type (in X) is finitely satisfiable in $((\mathfrak{A},\sigma),(\mathfrak{B},\tau),\bar{a},\bar{U}^+,\bar{U}^-,\bar{b},\bar{V}^+,V,\bar{V}^-)$

$$\{c_i \in X\} \cup \{\varphi(c_o,\ldots,c_{n-1},C_o^+,\ldots,C_{r-1}^+,X,C_o^-,\ldots,C_{s-1}^-)$$
$$\rightarrow \varphi(d_o,\ldots,d_{n-1},D_o^+,\ldots,D_{r-1}^+,D,D_o^-,\ldots,D_{s-1}^-|\varphi \in \Phi\}.$$

Hence, it is realized, say by $U \in \sigma$. Then $(p^o,p^1 \cup \{(U,V)\},p^2)$ is an exten - sion in I with the desired properties.

4.8 <u>Exercise</u>. Show: $(\mathfrak{A},\sigma)$ and $(\mathfrak{B},\tau)$ are partially homeomorphic iff they are $(L_{\infty\omega})_t$-equivalent.

The finite approximations of $\simeq^t$ are the relations $\simeq_\xi^t$ of the next definition for finite ordinals ξ.

4.9 <u>Definition</u>. Let ξ be an ordinal. We write $(I_\eta)_{\eta < \xi}:(\mathfrak{A},\sigma) \simeq_\xi^t (\mathfrak{B},\tau)$ and say that $(\mathfrak{A},\sigma)$ and $(\mathfrak{B},\tau)$ are <u>ξ-partially homeomorphic</u> via $(I_\eta)_{\eta < \xi}$, if each I_η is a non-empty set of partial homeomorphisms and the following back and forth properties hold for any η' and η with $\eta' < \eta < \xi$:

(forth$_1$) For $p \in I_\eta$ and $a \in A$ there is $q \in I_{\eta'}$ with $p \subset q$ and $a \in dom(q^o)$.

(forth$_2$) For $p \in I_\eta$, $a \in dom(p^o)$ and $U \in \sigma$ with $a \in U$ there are $q \in I_\eta$ and $V \in \tau$ such that $p \subset q$, $p^o(a) \in V$ and $U \; q^2 \; V$.

(back$_1$) For $p \in I_\eta$ and $b \in B$ there is $q \in I_{\eta'}$ with $p \subset q$ and $b \in rg(q^o)$.

(back$_2$) For $p \in I_\eta$, $b \in rg(q^o)$, say $p^o(a) = b$, and $V \in \tau$ with $b \in V$ there are $q \in I_{\eta'}$ and $U \in \sigma$ such that $p \subset q$, $a \in U$ and $U \; q^1 \; V$.

We write $(\mathfrak{A},\sigma) \simeq^t_\xi ((\mathfrak{B},\tau)$ if there is such a sequence $(I_\eta)_{\eta < \xi}$.

Note that any two structures are 1-partially homeomorphic. Our terminology is not standard: sometimes two structures are called ξ-homeomorphic, if they are $(\xi + 1)$-homeomorphic in our sense. - The name "finite approximation" is justified by the next lemma, whose proof is left to the reader.

4.10 Lemma. (a) If $(\mathfrak{A},\sigma) \simeq^t_p (\mathfrak{B},\tau)$ (in particular, if $(\mathfrak{A},\sigma) \simeq^t (\mathfrak{B},\tau)$), then $(\mathfrak{A},\sigma) \simeq^t_\xi (\mathfrak{B},\tau)$ for any ξ.

(b) If $(\mathfrak{A},\sigma) \simeq^t_{n+2} (\mathfrak{B},\tau)$ and A is a set of n elements, then $(\mathfrak{A},\sigma) \simeq^t (\mathfrak{B},\tau)$.

Note that for $\simeq^t_\xi$ the corresponding analogues of the results in 4.4 hold.

4.11 Exercise. Let L be one-sorted. Show that there is a many-sorted L' and, for $n \in \omega$, a set Φ_n of L'_t-sentences such that for all L-structures $(\mathfrak{A}_1,\sigma_1)$ and $(\mathfrak{A}_2,\sigma_2)$ we have:

$$(\mathfrak{A}_1,\sigma_1) \simeq^t_n (\mathfrak{A}_2,\sigma_2) \quad \text{iff} \quad ((\mathfrak{A}_1,\sigma_1),(\mathfrak{A}_2,\sigma_2),\ldots) \models \Phi_n$$

for an appropriate choice of the universes and relations in ...

Show that Φ_n and Φ_p of exercise 4.5 may be chosen such that $\Phi_0 \subset \Phi_1 \subset \ldots$ and $\Phi_p = \bigcup_{n \in \omega} \Phi_n$. In particular, if $((\mathfrak{A},\sigma),(\mathfrak{B},\tau))$ is recursively saturated and $(\mathfrak{A},\sigma) \simeq^t_n (\mathfrak{B},\tau)$ for $n \in \omega$, then $(\mathfrak{A},\sigma) \simeq^t_p (\mathfrak{B},\tau)$.

4.12 Exercise. Show that ω-partially homeomorphic structures are L_t-equivalent. (Hint: Argue as in the proof of 4.6 showing that each $p \in I_n$ preserves formulas of "rank" $\leq n$. Or, derive the result from 4.6 using the preceding exercise).

Our aim is to prove the converse of 4.12 for finite L. For that we show that the relations $\simeq_n^t$ are "definable" in L_t. <u>For the rest of the section let all similarity types be finite.</u>

Given $k \in \omega$ let Ψ_k be the finite set

$$\Psi_k = \{\psi \in L_{\omega\omega} \mid \psi = \psi(w_o,\dots,w_{k-1}) \text{ and } \psi \text{ is of the form } Rx_o\dots x_{n-1}, x_o = x_1 \text{ or } fx_o\dots x_{n-1} = x_n\}.$$

Recall that $w_o, w_1, \dots$ and $W_o, W_1, \dots$ are the variables resp. set variables. For $i \in \omega$, let $X_i = W_{2i}$ and $Y_i = W_{2i+1}$.

Given $(\mathfrak{A},\sigma), a_o,\dots,a_{k-1} \in A, U_o^+,\dots,U_{r-1}^+ \in \sigma$, $U_o^-,\dots,U_{s-1}^- \in \sigma$ define $\varphi^o_{\bar a,\bar U^+,\bar U^-}$ by

$$\varphi^o_{\bar a,\bar U^+,\bar U^-} = \bigwedge_{\substack{\psi \in \Psi_k \\ \mathfrak{A} \models \psi[\bar a]}} \psi \wedge \bigwedge_{\substack{\psi \in \Psi_k \\ \mathfrak{A} \models \neg\psi[\bar a]}} \neg\psi \wedge \bigwedge_{\substack{(\mathfrak{A},\sigma) \models \psi[\bar a,\bar U^+,\bar U^-] \\ \psi = v_i \in X_j \text{ or } \psi = \neg v_i \in Y_j}} \psi$$

Then, for any $(\mathfrak{B},\tau), b_o,\dots,b_{k-1} \in B$, $V_o^+,\dots,V_{r-1}^+, V_o^-,\dots,V_{s-1}^- \in \tau$ we have

$(\mathfrak{B},\tau) \models \varphi^o_{\bar a,\bar U^+,\bar U^-}[\bar b,\bar V^+,\bar V^-]$ iff $(\{(a_i,b_i) \mid i < k\}$, $\{(U_i^+,V_i^+) \mid i < r\}, \{(U_i^-,V_i^-) \mid i < s\})$ is a partial homeomorphism.

For $n > 0$ we define $\varphi^n_{\bar a,\bar U^+,\bar U^-}$ by

$$\varphi^n_{\bar a,\bar U^+,\bar U^-} = \underbrace{\bigwedge_{a \in A} \exists w_k\, \varphi^{n-1}_{\bar a a,\bar U^+,\bar U^-}}_{(\text{forth}_1)} \wedge \underbrace{\bigwedge_{0 \le i < k} \bigwedge_{\substack{U \in \sigma \\ a_i \in U}} \exists Y_s \ni w_i\, \varphi^{n-1}_{\bar a,\bar U^+,\bar U^- U}}_{(\text{forth}_2)}$$

$$\wedge \underbrace{\forall w_k \bigvee_{a \in A} \varphi^{n-1}_{\bar a a,\bar U^+,\bar U^-}}_{(\text{back}_1)} \wedge \underbrace{\bigwedge_{0 \le i < k} \forall X_r \ni w_i \bigvee_{\substack{U \in \sigma \\ a_i \in U}} \varphi^{n-1}_{\bar a,\bar U^+ U,\bar U^-}}_{(\text{back}_2)} .$$

Note that all conjunctions and disjunctions are finite, since one can show by induction on n that, for fixed k,r,s there are only finitely many formulas of the form $\varphi^n_{\bar a,\bar U^+,\bar U^-}$.

Each $\varphi^n_{\bar a,\bar U^+,\bar U^-}$ is an L_t-formula. In case $k = r = s = 0$ denote $\varphi^n_{\emptyset,\emptyset,\emptyset}$ by $\varphi^n_{(\mathfrak{A},\sigma)}$. $\varphi^n_{(\mathfrak{A},\sigma)}$ is an L_t-sentence.

Given $(\mathfrak{A},\sigma)$ and $(\mathfrak{B},\tau)$ put

$$I_n = \{p \mid p \text{ is of the form } (\{(a_i,b_i) \mid i < k\},\{(U_i^+,V_i^+) \mid i < r\},\{(U_i^-,V_i^-) \mid i < s\})$$
$$\text{and } (\mathfrak{B},\tau) \models \varphi^n_{\bar a,\bar U^+,\bar U^-}[\bar b,\bar V^+,\bar V^-]\} .$$

The sequence $(I_n)_{n<\omega}$ has the back and forth properties listed in 4.9; for each property we have specified in the definition of $\varphi^n_{\bar a,\bar U^+,\bar U^-}$ what part of $\varphi^n_{\bar a,\bar U^+,\bar U^-}$ is needed.

If $(\mathfrak{B},\tau) \models \varphi^n_{(\mathfrak{A},\sigma)}$ then I_n is non-empty (since $(\emptyset,\emptyset,\emptyset) \in I_n$).

Thus, if $(\mathfrak{A},\sigma) \equiv^t (\mathfrak{B},\tau)$ then $(I_n)_{n<\omega}: (\mathfrak{A},\sigma) \simeq^t_\omega (\mathfrak{B},\tau)$.

In particular, we have proved (compare 4.12):

4.13 <u>Ehrenfeucht-Fraïssé theorem</u>. Let L be finite. For any two topological structures $(\mathfrak{A},\sigma)$ and $(\mathfrak{B},\tau)$ the following are equivalent:

(i) $(\mathfrak{A},\sigma)$ and $(\mathfrak{B},\tau)$ are L_t-equivalent.

(ii) For all $n \in \omega$, $(\mathfrak{B},\tau) \models \varphi^n_{(\mathfrak{A},\sigma)}$.

(iii) $(\mathfrak{A},\sigma)$ and $(\mathfrak{B},\tau)$ are ω-partially homeomorphic.

An analysis of the proof of 4.6 shows that $(\mathfrak{B},\tau) \models \varphi^n_{(\mathfrak{A},\sigma)}$ holds if there is a partial homeomorphism that can be extended back and forth n-times. Hence

$$(\mathfrak{B},\tau) \models \varphi^n_{(\mathfrak{A},\sigma)} \quad \text{iff} \quad (\mathfrak{A},\sigma) \simeq^t_{n+1} (\mathfrak{B},\tau).$$

In particular by 4.10b

4.14 <u>Theorem</u>. Each finite topological structure may be characterized up to homeomorphism by an L_t-sentence.

4.15 <u>Exercise</u>. Show the equivalence of

(i) $(\mathfrak{A},\sigma) \simeq^t_{n+1} (\mathfrak{B},\tau)$.

(ii) $(\mathfrak{B},\tau) \models \varphi^n_{(\mathfrak{A},\sigma)}$.

(iii) $(\mathfrak{A},\sigma)$ and $(\mathfrak{B},\tau)$ satisfy the same L_t-sentences of rank $\leq n$ (where the definition of the rank of a formula is the natural extension to L_t of the definition in [6]).

Often when applying the back and forth methods, the following lemma will be useful.

4.16 <u>Convergence lemma</u>. Let L, L^1 and L^2 be such that $L \subset L^1 \cap L^2$. Suppose $(\mathfrak{B},\tau)$ is an L^2-structure. Assume that for each $n \in \omega$, there is an L^1-structure $(\mathfrak{A}_n,\sigma_n)$ such that $(\mathfrak{A}_n \upharpoonright L,\sigma_n) \simeq^t_n (\mathfrak{B} \upharpoonright L,\tau)$. Then there are an L^1-structure $(\mathfrak{A}^*,\sigma^*)$ and an L^2-structure $(\mathfrak{B}^*,\tau^*)$ such that

(1) $(\mathfrak{A}^* \upharpoonright L,\tilde{\sigma}^*) = (\mathfrak{B}^* \upharpoonright L,\tilde{\tau}^*)$

(2) $A^* \cup \sigma^* \cup B^* \cup \tau^*$ is denumerable.

(3) $(\mathfrak{B},\tau) \equiv (\mathfrak{B}^*,\tau^*)$, i.e. $(\mathfrak{B},\tau)$ and $(\mathfrak{B}^*,\tau^*)$ are L^2_2-equivalent.
In particular, $(\mathfrak{B},\tau) \equiv^t (\mathfrak{B}^*,\tilde{\tau}^*)$.

(4) $(\mathfrak{A}^*,\sigma^*)$ is a model of each L^1_2-sentence holding in all $(\mathfrak{A}_n,\sigma_n)$. In particular $(\mathfrak{A}^*,\tilde{\sigma}^*)$ is a model of each L^1_t-sentence holding in all $(\mathfrak{A}_n,\sigma_n)$.

<u>Proof</u>. Clearly it suffices to show the existence of $(\mathfrak{A}^*,\sigma^*)$ and $(\mathfrak{B}^*,\tau^*)$ with homeomorphic L-reducts instead of (1). We introduce a many-sorted similarity type which enables us to speak of structures of the form $((\mathfrak{A}',\sigma'),(\mathfrak{B}',\tau'))$ where $(\mathfrak{A}',\sigma')$ and $(\mathfrak{B}',\tau')$ are L^1 resp. L^2-structures. For $\varphi \in L^1_2$ resp. $\psi \in L^2_2$ let φ^1 resp. ψ^2 denote a sentence of this many-sorted language such that

$$((\mathfrak{A}',\sigma'),(\mathfrak{B}',\tau')) \models \varphi^1 \quad \text{iff} \quad (\mathfrak{A}',\sigma') \models \varphi ,$$

$$((\mathfrak{A}',\sigma'),(\mathfrak{B}',\tau')) \models \psi^2 \quad \text{iff} \quad (\mathfrak{B}',\tau') \models \psi.$$

By assumption, each finite subset of

$$\Phi = \{(\varphi^n_{(\mathfrak{B}\upharpoonright L,\tau)})^1 \mid n \in \omega\} \cup \{\varphi^1 \mid \varphi \in L^1_2 \text{ and } (\mathfrak{A}_n,\sigma_n) \models \varphi \text{ for all } n\}.$$
$$\cup \{\psi^2 \mid \psi \in L^2_2,\ (\mathfrak{B},\tau) \models \psi\}$$

is satisfiable. Thus there is a denumerable recursively saturated model $((\mathfrak{A}^*,\sigma^*),(\mathfrak{B}^*,\tau^*))$ of Φ. $(\mathfrak{A}^*,\sigma^*)$ and $(\mathfrak{B}^*,\tau^*)$ satisfy (2),(3),(4),and by 4.13, $(\mathfrak{A}^*,\sigma^*) \equiv^t (\mathfrak{B}^*,\tau^*)$. But then, by 4.7, $(\mathfrak{A}^*,\sigma^*)$ and $(\mathfrak{B}^*,\tau^*)$ are homeomorphic.

4.17 Corollary. Suppose the topological structures $(\mathfrak{A},\sigma)$ and $(\mathfrak{B},\tau)$ are L_t-equivalent. Then there is a topological structure $(\mathfrak{A}^*,\sigma^*)$ and there are bases σ_1 and σ_2 of σ^* such that $(\mathfrak{A}^*,\sigma_1) \equiv (\mathfrak{A},\sigma)$ and $(\mathfrak{A}^*,\sigma_2) \equiv (\mathfrak{B},\tau)$.

Proof. For each $n \in \omega$, take $(\mathfrak{A},\sigma)$ as $(\mathfrak{A}_n,\sigma_n)$ in the convergence lemma.

We close this section with two applications.First we prove the analogue of the Keisler-Shelah ultraproduct theorem.

Given an ultrafilter D over a set I, $(\mathfrak{A},\sigma)^I/_D$ the ultrapower of $(\mathfrak{A},\sigma)$ is in a natural way a weak structure. If $(\mathfrak{A},\sigma) \models$ bas then, by Łos theorem, $(\mathfrak{A},\sigma)^I/_D \models$ bas.

4.18 Theorem. Two topological structures are L_t-equivalent iff they have homeomorphic ultrapowers.

Proof. First suppose that the ultrapowers $(\mathfrak{A}_1,\sigma_1)^{I_1}/_{D_1}$ and $(\mathfrak{A}_2,\sigma_2)^{I_2}/_{D_2}$ are homeomorphic. Then

$(\mathfrak{A}_1,\sigma_1)^{I_1}/_{D_1} \equiv^t (\mathfrak{A}_2,\sigma_2)^{I_2}/_{D_2}$ and, by Łos theorem,

$(\mathfrak{A}_j,\sigma_j)^{I_j}/_{D_j} \equiv (\mathfrak{A}_j,\sigma_j)$ for $j = 1,2$. Hence $(\mathfrak{A}_1,\sigma_1) \equiv^t (\mathfrak{A}_2,\sigma_2)$. -

Now assume that $(\mathfrak{A}_1,\sigma_1) \equiv^t (\mathfrak{A}_2,\sigma_2)$. By the preceding corollary there are $\mathfrak{B},\tau_1$ and τ_2 such that $\tilde{\tau}_1 = \tilde{\tau}_2$ and $(\mathfrak{A}_j,\sigma_j) \equiv (\mathfrak{B},\tau_j)$ for $j = 1,2$. By the Keisler-Shelah-theorem we find an ultrafilter D over a set I such that

$$(\mathfrak{A}_j,\sigma_j)^I/_D \simeq (\mathfrak{B},\tau_j)^I/_D \quad \text{for } j = 1,2.$$

Put $(\mathfrak{B}^*,\tau_j^*) = (\mathfrak{B},\tau_j)^I/_D$. Since $(\mathfrak{B},\tilde{\tau}_1) = (\mathfrak{B},\tilde{\tau}_2)$, we have $(\mathfrak{B}^*,\tilde{\tau}_1^*) = (\mathfrak{B}^*,\tilde{\tau}_2^*)$. Hence

$$(\mathfrak{A}_1,\sigma_1)^I/_D \simeq (\mathfrak{B}^*,\tau_1^*) \simeq^t (\mathfrak{B}^*,\tau_2^*) \simeq (\mathfrak{A}_2,\sigma_2)^I/_D.$$

Finally, we derive the following

4.19 Theorem. Each L_2-sentence invariant for topologies is equivalent in topological structures to an L_t-sentence, i.e. if $\varphi \in L_2$ is invariant for topologies then there is $\psi \in L_t$ with $\models \varphi \leftrightarrow \psi$. Moreover we may choose for ψ a disjunction of some $\varphi^n_{(\mathfrak{A},\sigma)}$.

We get this theorem from the following lemma for $\Phi = \{\underline{bas}\}$. For $\Phi = \emptyset$ we obtain that each invariant L_2-sentence is equivalent to an L_t-sentence.

4.20 Lemma. Let $\Phi \cup \{\varphi\}$ be a set of L_2-sentences. Suppose that φ is invariant for models of Φ, i.e. whenever $(\mathfrak{A},\sigma) \models \Phi$, $(\mathfrak{A},\tau) \models \Phi$ and $\tilde{\sigma} = \tilde{\tau}$ then

$$(\mathfrak{A},\sigma) \models \varphi \quad \text{implies} \quad (\mathfrak{A},\tau) \models \varphi.$$

Then there is an L_t-sentence ψ with $\Phi \models \varphi \leftrightarrow \psi$.

Proof. If $\Phi \cup \{\varphi\}$ is not satisfiable let ψ be $\exists x \neg x = x$. Otherwise, for $n \in \omega$ put

$$\psi^n = \bigvee\{\varphi^n_{(\mathfrak{A},\sigma)} \mid (\mathfrak{A},\sigma) \models \Phi \cup \{\varphi\}\}.$$

Then ψ^n is an L_t-sentence and

(1) $\Phi \models \varphi \rightarrow \psi^n$ (2) $\models \psi^{n+1} \rightarrow \psi^n$.

It suffices to show that $\Phi \cup \{\psi^n \mid n \in \omega\} \models \varphi$; because then, by compactness, (1) and (2), we have $\Phi \models \varphi \leftrightarrow \psi^n$ for some n.

So, suppose $(\mathfrak{B},\tau) \models \Phi \cup \{\psi^n \mid n \in \omega\}$. Since $(\mathfrak{B},\tau) \models \psi^n$ there is $(\mathfrak{A}_n,\sigma_n)$ such that $(\mathfrak{A}_n,\sigma_n) \models \Phi \cup \{\varphi\}$, and $(\mathfrak{B},\tau) \models \varphi^n_{(\mathfrak{A}_n,\sigma_n)}$, i.e. $(\mathfrak{A}_n,\sigma_n) \simeq^t_n (\mathfrak{B},\tau)$. By the convergence lemma, we find $\mathfrak{A}^*,\sigma^*$ and τ^* such that $\tilde{\sigma}^* = \tilde{\tau}^*$, $(\mathfrak{A}^*,\sigma^*) \models \Phi \cup \{\varphi\}$ and $(\mathfrak{A}^*,\tau^*) \equiv (\mathfrak{B},\tau)$. In particular $(\mathfrak{A}^*,\tau^*) \models \Phi$. But then, by Φ-invariance of φ, we have $(\mathfrak{A}^*,\tau^*) \models \varphi$, i.e. $(\mathfrak{B},\tau) \models \varphi$.

Taking for example $\Phi = \{\underline{bas} \wedge \underline{haus}\}$, we see that each L_2-sentence invariant for Hausdorff spaces is equivalent in Hausdorff spaces to an L_t-sentence.

Since, for infinite L, $\varphi^n_{(\mathfrak{A},\sigma)}$ is not a finite formula the Ehrenfeucht-Fraïssé theorem does not hold in the form 4.13 for infinite L. Nevertheless 4.16 and the ultrapower theorem are still true. The following exercise shows how to derive them with the methods introduced above.

4.21 Exercise. Let L be infinite, countable or uncountable.

(a) Suppose $(\mathfrak{A},\sigma) \equiv^t (\mathfrak{B},\tau)$. Show that there are homeomorphic structures $(\mathfrak{A}',\sigma')$ and $(\mathfrak{B}',\tau')$ such that $(\mathfrak{A},\sigma) \equiv (\mathfrak{A}',\sigma')$ and $(\mathfrak{B},\tau) \equiv (\mathfrak{B}',\tau')$. (Hint: By a compactness argument it suffices to show the statement for finite L).

(b) Show: $(\mathfrak{A},\sigma) \equiv^t (\mathfrak{B},\tau)$ iff for each finite $L' \subset L$ and each $n \in \omega$ $(\mathfrak{A} \upharpoonright L',\sigma) \simeq^t_n (\mathfrak{B} \upharpoonright L',\tau)$.

(c) Prove the following convergence lemma: Let L, L^1 and L^2 be such that $L \subset L^1 \cap L^2$. Suppose $(\mathfrak{B},\tau)$ is an L^2-structure, and that for each $n \in \omega$, and each finite $L' \subset L$ there is an L^1-structure $(\mathfrak{A}_{(n,L')},\sigma_{(n,L')})$ such that $(\mathfrak{A}_{(n,L')}\upharpoonright L',\sigma_{(n,L')}) \simeq^t_n (\mathfrak{B}\upharpoonright L',\tau)$. Then there are an L^1-structure $(\mathfrak{A}^*,\sigma^*)$ and an L^2-structure $(\mathfrak{B}^*,\tau^*)$ such that

(i) $(\mathfrak{A}^*\upharpoonright L,\tilde{\sigma}^*) = (\mathfrak{B}^*\upharpoonright L,\tilde{\tau}^*)$

(ii) $(\mathfrak{B}^*,\tau^*) \equiv (\mathfrak{B},\tau)$

(iii) If $\varphi \in L^1_2$, say $\varphi \in L'_2$ where L' is finite, and if for each n, $(\mathfrak{A}_{(n,L')},\sigma_{(n,L')})$ is a model of φ, then $(\mathfrak{A}^*,\sigma^*) \models \varphi$.

§ 5 Interpolation and preservation

We prove in this section the interpolation theorem for L_t and derive preservation theorems for some relations between topological structures. We obtain the results applying the back and forth methods of the preceding section. From now on, unless otherwise stated, $(\mathfrak{A},\sigma),(\mathfrak{B},\tau),\ldots$ will denote topological structures.

5.1 Interpolation theorem. For $i = 1,2$ let $\varphi_i \in L^i_t$. Put $L = L^1 \cap L^2$.

If $\models_t \varphi_1 \to \varphi_2$ then for some $\psi \in L_t$, $\models_t \varphi_1 \to \psi$ and $\models_t \psi \to \varphi_2$.

Proof. Note that here and also in the next theorems we may assume that the similarity types are finite. - Put $\psi = \exists x \neg x = x$, if φ_1 has no topological model. Otherwise, let for $n \in \omega$,

$$\psi^n = \bigvee\{\varphi^n_{(\mathfrak{A}\upharpoonright L,\sigma)} \mid (\mathfrak{A},\sigma) \models \varphi_1,\ (\mathfrak{A},\sigma)L^1\text{-structure}\}.$$

Since $\models_t \varphi_1 \to \psi^n$ and $\models_t \psi^{n+1} \to \psi^n$, it suffices to show that $\{\psi^n \mid n \in \omega\} \models_t \varphi_2$. So suppose $(\mathfrak{B},\tau)$ is an L^2-structure and a model of $\{\psi^n \mid n \in \omega\}$. Then for each n, there is an L^1-structure $(\mathfrak{A}_n,\sigma_n)$ with

$$(\mathfrak{A}_n,\sigma_n) \models \varphi_1 \quad\text{and}\quad (\mathfrak{B},\tau) \models \varphi^n_{(\mathfrak{A}_n\upharpoonright L,\sigma_n)} .$$

By the convergence lemma 4.16 there are an L^1-structure $(\mathfrak{A}^*,\sigma^*)$, and an

L^2-structure $(\mathfrak{B}^*,\tau^*)$ such that

$$(\mathfrak{A}^*\upharpoonright L,\sigma^*) = (\mathfrak{B}^*\upharpoonright L,\tau^*),\ (\mathfrak{A}^*,\sigma^*) \models \varphi_1 \quad \text{and} \quad (\mathfrak{B}^*,\tau^*) \equiv^t (\mathfrak{B},\tau).$$

But then $(\mathfrak{B}^*,\tau^*,(k^{\mathfrak{A}^*})_{k \in L^1 - L})$ is a model of φ_1 and therefore of φ_2. Hence $(\mathfrak{B},\tau) \models \varphi_2$.

We remark that 8.8.4 contains a syntactic proof of the interpolation theorem.

5.2 <u>Exercise</u>. Let $L = L^1 \cap L^2$, $\varphi_1 \in L^1{}_2$, $\varphi_2 \in L^2{}_2$. Assume that φ_2 is invariant. Show that the following are equivalent:

(i) $\models \varphi_1 \to \varphi_2$.

(ii) There is a $\psi \in L_t$ such that $\models \varphi_1 \to \psi$ and $\models \psi \to \varphi_2$.

Considering the back and forth properties for each sort, it is possible to introduce notions like "partial homeomorphic" also for many-sorted structures, and to derive the corresponding results. For example one gets for many-sorted L:

(1) Each L_2-sentence invariant for topologies is equivalent in topological structures to an L_t-sentence.

(2) Suppose $\models_t \varphi_1 \to \varphi_2$ where $\varphi_1,\varphi_2 \in L_t$. Then there is a $\psi \in L_t$ with

a) $\models_t \varphi_1 \to \psi$ and $\models_t \psi \to \varphi_2$.

b) Each relation and function symbol in ψ occurs in φ_1 and φ_2 .

c) If ψ contains a term of sort i, then so do φ_1 and φ_2 (τ "truth" is needed in case φ_1 and φ_2 have no common sort).

d) If ψ contains a set variable of sort i, then so do φ_1 and φ_2.

Note that from the many-sorted interpolation theorem (2) one obtains (1) using the technique of additional universes sketched in 4.5. In particular a syntactic proof of the many-sorted interpolation theorem (see 8.8.4) yields a syntactic proof of the fact that the invariant L_2-sentences are the L_t-sentences.

5.3 <u>Exercise</u>. Characterize the invariant sentences for structures of type $(\mathfrak{A},\sigma_1,\ldots,\sigma_n)$ where $\sigma_1,\ldots,\sigma_n$ are topologies on A. Derive the corresponding interpolation theorem.

Now we apply the methods of the preceding section to give a uniform treatment of some preservation theorems.

5.4 Definition. $(\mathfrak{A},\sigma)$ is a substructure of $(\mathfrak{B},\tau)$, in symbols $(\mathfrak{A},\sigma) \subset (\mathfrak{B},\tau)$, if $\mathfrak{A}$ is a substructure of $\mathfrak{B}$ (in the algebraic sense) and σ is the topology on A induced by τ. If in addition A is a dense subset of B, we call $(\mathfrak{A},\sigma)$ a dense substructure of $(\mathfrak{B},\tau)$, $(\mathfrak{A},\sigma) \subset_d (\mathfrak{B},\tau)$. If A is an open subset of B, i.e. $A \in \tau$, we call $(\mathfrak{A},\sigma)$ an open substructure of $(\mathfrak{B},\tau)$, $(\mathfrak{A},\sigma) \subset_o (\mathfrak{B},\tau)$. - $(\mathfrak{B},\tau)$ is then called an extension, resp. dense extension, resp. open extension of $(\mathfrak{A},\sigma)$.

First we characterize those L_t-sentences φ which are preserved under extensions, i.e. satisfying:

$$(\mathfrak{A},\sigma) \models \varphi \text{ and } (\mathfrak{A},\sigma) \subset (\mathfrak{B},\tau) \text{ implies } (\mathfrak{B},\tau) \models \varphi .$$

5.5 Definition. An L_t-formula is existential (universal) iff it is in negation normal form and does not contain any universally (existentially) quantified individual variable.

A simple proof shows that every existential (universal) sentence is preserved under extensions (substructures). haus and reg are universal sentences.

Let the binary relation E on weak models of bas be defined by:

$$(\mathfrak{A},\sigma) \, E \, (\mathfrak{B},\tau) \text{ iff } (\mathfrak{A},\tilde{\sigma}) \text{ is homeomorphic to a substructure of } (\mathfrak{B},\tilde{\tau}).$$

Then any $\varphi \in L_t$ is preserved under extensions iff it is preserved under E.

Note that for any weak structures, $(\mathfrak{A},\sigma) \, E \, (\mathfrak{B},\tau)$ holds iff there are a map π^o and relations $\pi^1, \pi^2 \subset \sigma \times \tau$ satisfying the properties (1) - (4) listed at the beginning of § 4, where instead of (1) we only require that π^o is an isomorphism of $\mathfrak{A}$ onto a substructure of $\mathfrak{B}$. Therefore, we obtain a countable approximation E_p and finite approximations E_n of E from $\simeq_p^t$ and $\simeq_n^t$, if we drop the condition (back$_1$). On countable models, E_p coincides with E. The relations E_n are "definable": the formulas $\psi^n_{\bar{a},\bar{U}^+,\bar{U}^-}$ corresponding to $\varphi^n_{\bar{a},\bar{U}^+,\bar{U}^-}$ are obtained from $\varphi^n_{\bar{a},\bar{U}^+,\bar{U}^-}$ by dropping the conjuncts that formalize the ($back_1$)-property, hence they are existential. - But each sentence preserved by E is equivalent to a disjunction of some of these existential sentences $\psi^n_{(\mathfrak{A},\sigma)}$ (the proof is similar to that of 4.19 and uses the corresponding generalization of 4.11).

5.6 Theorem. $\varphi \in L_t$ is preserved under extensions iff there is an existential sentence $\psi \in L_t$ with $\models_t \varphi \leftrightarrow \psi$.

Since the negation of an existential sentence is equivalent to an universal sentence, we have:

5.7 Corollary. $\varphi \in L_t$ is preserved under substructures iff there is a universal $\psi \in L_t$ with $\models_t \varphi \leftrightarrow \psi$.

Let $\varphi = \varphi(X) \in L_t$ be positive in X. Then $\exists x\, \varphi(\{x\})$ (i.e. the L_t-sentence $\exists x \chi$, where χ is obtained from φ replacing each subformula $t \in X$ by $t = x$) is preserved under substructures. By 5.7 there is a universal ψ equivalent to $\exists x\, \varphi(\{x\})$. Find such a ψ!

5.8 Exercise. 1) Let $L = L^1 \cap L^2$, $\varphi_1 \in L^1{}_t$, $\varphi_2 \in L^2{}_t$. Assume that $L^2 - L$ contains no function symbols. Show that the following are equivalent.

(i) For any topological $(L^1 \cup L^2)$- structures $(\mathfrak{A},\sigma),(\mathfrak{B},\tau)$:
$(\mathfrak{A},\sigma) \models \varphi_1$ and $(\mathfrak{A},\sigma) \subset (\mathfrak{B},\tau)$ imply $(\mathfrak{B},\tau) \models \varphi_2$.

(ii) There is an existential $\psi \in L_t$ with $\models_t \varphi_1 \rightarrow \psi$ and $\models_t \psi \rightarrow \varphi_2$.

2) Suppose that the class of topological models of an L_t-theory T is closed under extensions. Show that there is a set T* of existential L_t-sentences with the same models. (Hint: Take as T* the set $\{\psi^n_{(\mathfrak{A}\restriction L',\sigma)} \mid n \in \omega, L' \subset L$ finite, $(\mathfrak{A},\sigma) \models T\}$ and argue, in case of infinite L, as in 4.21).

3) Relativize 5.6 to models of a given theory T, i.e. show: if $(\mathfrak{A},\sigma) \models T \cup \{\varphi\}$, $(\mathfrak{A},\sigma) \subset (\mathfrak{B},\tau)$ and $(\mathfrak{B},\tau) \models T$ imply $(\mathfrak{B},\tau) \models \varphi$, then there is an existential ψ with $T \models_t \varphi \leftrightarrow \psi$.

Now we characterize the sentences preserved under dense extensions. For example, for unary P, the sentence

$$(+) \qquad \forall x\, \forall X \ni x\, \exists y\, (Py \wedge y \in X).$$

expressing that P is dense is preserved under dense extensions.

5.9 <u>Definition</u>. An L_t-formula φ is <u>d-existential</u> iff it is in negation normal form and each subformula of φ beginning with a universally quantified individual variable is of the form $\forall x\ \forall X \ni x\psi$ where ψ does not contain x free.

Thus (+) is d-existential. A simple induction shows that any d-existential sentence is preserved under dense extensions. Define the binary relation D by:

for any weak models $(\mathfrak{A},\sigma)$ and $(\mathfrak{B},\tau)$ of <u>bas</u>

$(\mathfrak{A},\sigma)$ D $(\mathfrak{B},\tau)$ iff $(\mathfrak{A},\tilde{\sigma})$ is homeomorphic to a dense substructure of $(\mathfrak{B},\tilde{\tau})$.

Note that $(\mathfrak{A},\sigma)$ D $(\mathfrak{B},\tau)$ hold iff there are π^o,π^1 and π^2 satisfying (1) - (4) of the beginning of § 4, but where π^o is now an isomorphism onto a dense substructure. Therefore to get a countable approximation D_p of D, we have to drop (back_1) in the definition of $\simeq_p^t$ and to insert the following condition (back_d) guaranteeing the density:

(back_d) For $p \in I$, $b \in B$ and $V \in \tau$ with $b \in V$ there are $q \in I$ and $b' \in V$ such that $p \subset q$ and $b' \in rg(q)$.

Show that D_p coincides with D on countable structures. - To "describe" the corresponding finite approximations D_n we introduce formulas $\chi^n_{\bar{a},\bar{U}^+,\bar{U}^-}$. $\chi^n_{\bar{a},\bar{U}^+,\bar{U}^-}$ is obtained from $\varphi^n_{\bar{a},\bar{U}^+,\bar{U}^-}$ dropping the part that formalizes the (back_1)-property but inserting the following conjunct corresponding to (back_d)

(*) $$\forall w_k\ \forall X_r \ni w_k\ \exists w_k\ (w_k \in X_r \wedge \bigvee_{a \in A} \chi^{n-1}_{\bar{a}a,\bar{U}^+,\bar{U}^-}).$$

Hence $\chi^n_{\bar{a},\bar{U}^+,\bar{U}^-}$ is d-existential. Therefore:

5.10 <u>Theorem</u>. A sentence $\varphi \in L_t$ is preserved under dense extensions iff there is a d-existential sentence $\psi \in L_t$ with $\models_t \varphi \leftrightarrow \psi$.

For a formula $\varphi(x,\bar{x},\bar{X}^+,\bar{Y}^-)$ and any new variable X, the formula $(\forall x\ \forall X \ni x\ \exists x(x \in X \wedge \varphi)$ expresses that the set $\{x|\varphi\}$ is dense, hence we denote this formula by "$\{x|\varphi\}$". Note that by (*), we can assume that in any d-existential formula any universal quantifier binding an individual variable occurs in the "prefix" of a subformula of type "$\{x|\varphi\}$".

5.11 Definition. An L_t-formula φ is a Σ-formula iff it is in negation normal form and each subformula of φ beginning with a universally quantified individual variable has the form $\forall x(x \in Y \rightarrow \psi)$ (abbreviated by $\forall x \in Y\psi$).

Note that each Σ-sentence is preserved under open extensions. When studying the relation of being an open extension we have instead of (back_1) the condition (back_0),

(back_0) For all $p \in I$, all V such that $U \; p^2 \; V$ for some U, and all $b \in V$ there is $q \in I$ with $p \subset q$ and $b \in rg(q^0)$.

Thus the corresponding formulas $\vartheta^{n}_{\bar{a},\bar{U}^+,\bar{U}^-}$ contain a conjunct of the form

$$\bigwedge_{o \leq i < s} \forall w_k \in Y_i \bigvee_{a \in A} \vartheta^{n-1}_{\bar{a}a,\bar{U}^+,\bar{U}^-} ,$$

hence they are Σ-formulas.

5.12 Theorem. An L_t-sentence φ is preserved under open extensions iff there is a Σ-sentence $\psi \in L_t$ with $\models_t \varphi \leftrightarrow \psi$.

In 8.8.7 we show that 5.12 and the corresponding theorem for end-extensions in classical model theory have a common generalization.

5.13 Exercise. A formula in negation normal form is called a Π-formula iff each subformula of φ beginning with an existentially quantified individual variable has the form $\exists x(x \in Y \wedge \psi)$ (abbreviated by $\exists x \in Y\psi$). A formula which is both, Π and Σ, is called a Δ-formula. - For a Π-formula φ and a new set variable Y let φ^Y be the L_t-formula obtained from φ restricting all universal individual quantifiers to Y, i.e. replacing each subformula $\forall x \chi$ by $\forall x \in Y \chi^Y$. Thus $\exists Y \ni t \varphi^Y$ is a Δ-formula. Suppose that L contains no function symbols besides c. Then given $(\mathfrak{A},\sigma)$ and a $U \in \sigma$ containing c^A there is an open substructure, denoted by $(\mathfrak{A},\sigma)\restriction U$, with universe U.

1) Show that for given $(\mathfrak{A},\sigma)$ and $(\mathfrak{B},\tau)$ the following are equivalent.

(i) There are $(\mathfrak{A}^*,\sigma^*)$ and $(\mathfrak{B}^*,\tau^*)$ L_t-equivalent to $(\mathfrak{A},\sigma)$ and $(\mathfrak{B},\tau)$ respectively, which contain homeomorphic open substructures.

(ii) $(\mathfrak{A},\sigma)$ and $(\mathfrak{B},\tau)$ satisfy the same Δ-sentences.

(Hint: Using a compactness argument find $(\mathfrak{A}_1,\sigma_1)$ L_t-equivalent to $(\mathfrak{A},\sigma)$ containing an open substructure $(\mathfrak{C},\sigma')$ that satisfies all Π-sentences ψ such that $(\mathfrak{A},\sigma) \models \exists U \ni c\ \psi^U$. Then all Σ-sentences which hold in $(\mathfrak{C},\sigma')$ hold in $(\mathfrak{B},\tau)$).

2) For $\varphi \in L_t$ show the equivalence of (i) and (ii).

(i) φ holds in a topological structure iff it holds in all sufficiently small open substructures (i.e. $(\mathfrak{A},\sigma) \models \varphi$ iff for some $U \in \sigma$ with $c \in U$ and all $V \in \sigma$, $c \in V \subset U: (\mathfrak{A},\sigma)\restriction V \models \varphi$).

(ii) There is a Δ-sentence ψ such that $\models_t \varphi \leftrightarrow \psi$.

The back and forth method is useful for obtaining preservation theorems for other relations, two examples are contained in the following exercise.

5.14 Exercise. 1) (Finer topologies). An L_t-sentence φ is preserved under finer topologies, if for any structure $\mathfrak{A}$ and topologies σ and τ,

$$(\mathfrak{A},\sigma) \models \varphi \quad \text{and} \quad \sigma \subset \tau \quad \text{imply} \quad (\mathfrak{A},\tau) \models \varphi.$$

An L_t-formula in negation normal form is called set-existential, if it does not contain any universally quantified set variable. - haus and disc are set-existential. Show that the L_t-sentences preserved under finer topologies are just the sentences which are equivalent to a set-existential sentence.

2) (Continuous homomorphic images) Characterize the sentences preserved under continuous homomorphic images.

3) For dense extensions, open extensions, finer topologies and continuous homomorphic images derive the results corresponding to 5.8.1 - 5.8.3.

§ 6 Products and Sums

(preservation theorems continued)

In this section we generalize some results on products and sums to topological logic. At some points we assume that the reader is familiar with the definitions and results of [5]. Recall that unless otherwise noted, all structures will be topological structures.

We denote by $\prod_I(\mathfrak{A}_i,\sigma_i)$ the (topological) product of the structures $(\mathfrak{A}_i,\sigma_i)$

for $i \in I$, i.e. the structure $(\prod_I \mathfrak{A}_i, \sigma)$ consisting of $\prod_I \mathfrak{A}_i$, the "classical" product of the $\mathfrak{A}_i$ and the product topology σ. Let $\overset{o}{\prod_I} \sigma_i$ be the natural basis of σ,

$$\overset{o}{\prod_I} \sigma_i = \{\prod_I U_i | U_i \in \sigma_i,\ U_i = A_i \text{ almost everywhere}\}.$$

For L containing no function symbols, we denote by $\sum_I (\mathfrak{A}_i, \sigma_i)$ the (topological) sum (or free union) of the structures $(\mathfrak{A}_i, \sigma_i)$. It consists of the "classical" sum $\sum_I \mathfrak{A}_i$ and the sum topology. Let $\overset{o}{\sum_I} \sigma_i$ be the natural basis of this topology

$$\overset{o}{\sum_I} \sigma_i = \{\dot{\bigcup_I} U_i | U_i \in \sigma \text{ for all } i \in I\}.$$

Note that the topological product $\prod_I (\mathfrak{A}_i, \sigma_i)$ in general does not coincide with the "classical" product of the two-sorted structures $(\mathfrak{A}_i, \sigma_i)$. And contrary to the classical case, there are L_t-sentences (e.g. disc) that are preserved under finite products but not under arbitrary products. -

On the other hand some theorems and proof techniques generalize to the present case, e.g.

6.1 Theorem. If $(\mathfrak{A}_i, \sigma_i) \equiv^t (\mathfrak{B}_i, \tau_i)$ for each $i \in I$, then $\prod_I (\mathfrak{A}_i, \sigma_i) \equiv^t \prod_I (\mathfrak{B}_i, \tau_i)$ and $\sum_I (\mathfrak{A}_i, \sigma_i) \equiv^t \sum_I (\mathfrak{B}_i, \tau_i)$.

Proof. A winning strategy for player II in the Ehrenfeucht game for, say, the product structure is obtained by playing in each component according to a winning strategy. Note that by 4.4.4 we may assume that in the game for the i-th component player II chooses A_i (resp. B_i), if B_i (resp. A_i) is chosen by I.

Using a global strategy one can strengthen 6.1, thus, e.g. obtaining:

If I and J are both infinite, and $(\mathfrak{A}, \sigma) \equiv^t (\mathfrak{B}, \tau)$, then $(\mathfrak{A}, \sigma)^I \equiv^t (\mathfrak{B}, \tau)^J$ (where $(\mathfrak{A}, \sigma)^I$ is $\prod_I (\mathfrak{A}, \sigma)$).

But it is also possible to derive these results from the results in [5] using the following remarks.

Given a set I denote by $\mathfrak{P}(I)$ and $(\mathfrak{P}(I), Fin)$ the Boolean algebra

$(P(I),\cap,\cup,-,\leq,\emptyset,I)$ of all subsets of I resp. the structure $(P(I),\cap,\cup,-,\leq,\emptyset,I,Fin)$ where Fin is the set of finite subsets of I. Let L^B resp. $L^{B'}$ be the corresponding similarity type.

It is easily shown that in the terminology of [5] we have

(1) $(\prod_I \mathfrak{A}_i, \overset{o}{\prod_I} \sigma_i)$ is a relativized generalized product relative to $(\mathfrak{P}(I),Fin)$.

(2) $(\sum_I \mathfrak{A}_i, \overset{o}{\sum_I} \sigma_i)$ is a relativized generalized product relative to $\mathfrak{P}(I)$.

Therefore

6.2 Theorem. a) Given any sentence $\varphi \in L_t$ we can find a number $m \in \omega$ such that: whenever φ is true in the sum of the structures $(\mathfrak{A}_i,\sigma_i)$ there is a set $I_o \subset I$, having at most m elements, and such that φ is true in the sum $\sum_{I'} (\mathfrak{A}_i,\sigma_i)$ provided $I_o \subset I' \subset I$.

b) Any L_t-sentence preserved under finite sums is preserved under arbitrary sums.

c) If I and J are both infinite and $(\mathfrak{A},\sigma) \equiv^t (\mathfrak{B},\tau)$, then $\sum_I (\mathfrak{A},\sigma) \equiv^t \sum_J (\mathfrak{B},\tau)$.

Proof. By (2), a) and b) are special cases of the results in [5].

c): By 4.17 we find $\mathfrak{C}$ and bases σ' and τ' of the same topology, $\tilde{\sigma}' = \tilde{\tau}'$, such that $(\mathfrak{A},\sigma) \equiv (\mathfrak{C},\sigma')$ and $(\mathfrak{B},\tau) \equiv (\mathfrak{C},\tau')$.

Hence

$$\sum_I (\mathfrak{A},\sigma) \equiv \sum_I (\mathfrak{C},\sigma') \equiv^t \sum_I (\mathfrak{C},\tilde{\sigma}') \equiv^t \sum_I (\mathfrak{C},\tau') \equiv \sum_J (\mathfrak{C},\tau') \equiv \sum_J (\mathfrak{B},\tau) .$$

The general preservation theorem of [5] tells us that given $\varphi \in L_t$ we can find effectively $\psi_1,\ldots,\psi_m \in L_2$ and $\chi = \chi(y_1,\ldots,y_m) \in L^{B'}_{\omega\omega}$ such that

$$\prod_I (\mathfrak{A}_i,\sigma_i) \models \varphi \quad \text{iff} \quad (\mathfrak{P}(I),Fin) \models \chi[S(\psi_1),\ldots,S(\psi_m)],$$

where $S(\psi_j) = \{i \in I \mid (\mathfrak{A}_i,\sigma_i) \models \psi_j\}$.

The formulas $\psi_1,\ldots,\psi_m$ obtained in the proof in [5] in general do not belong to L_t. But we need L_t-formulas in order to carry over to the present case the decidability results of [5]. We show how to modify the proof of [5] to get ψ_i lying in L_t.

If $a_1,\dots,a_n \in \overset{o}{\Pi}A_i$ is denoted by $\bar{a}$, let $\bar{a}(i)$ be $a_1(i),\dots,a_n(i)$; similarly, if $U_1,\dots,U_n \in \overset{o}{\Pi}\sigma_i$ is denoted by $\bar{U}$, let $\bar{U}(i)$ be $U_1(i),\dots,U_n(i)$, where $U_j(i) = \{a(i) \mid a \in U_j\}$.

6.3 Theorem. Given $\varphi = \varphi(x_1,\dots,x_n,X_1^+,\dots,X_r^+,Y_1^-,\dots,Y_s^-) \in L_t$ we can find effectively $\psi_1,\dots,\psi_m \in L_t, \psi_i = \psi_i(\bar{x},\bar{X}^+,\bar{Y}^-)$, and $\chi \in L^{B'}_{\omega\omega}, \chi = \chi(y_1,\dots,y_m)$ such that φ is determined by $(\chi;\psi_1,\dots,\psi_m)$, i.e. given any system of structures $(\mathfrak{A}_i,\sigma_i)$, any $a_1,\dots,a_n \in \prod_I A_i$, any $U_1,\dots,U_r,V_1,\dots,V_s \in \overset{o}{\prod_I} \sigma_i$, we have

$$\prod_I (\mathfrak{A}_i,\sigma_i) \models \varphi[\bar{a},\bar{U},\bar{V}] \quad \text{iff} \quad (\mathfrak{P}(I),\mathrm{Fin}) \models \chi[S(\psi_1),\dots,S(\psi_m)],$$

where

$$S(\psi_j) = \{i \mid (\mathfrak{A}_i,\sigma_i) \models \psi_j[\bar{a}(i),\bar{U}(i),\bar{V}(i)]\} .$$

Moreover χ may be chosen "monotomic", i.e. such that

$$(\mathfrak{P}(I),\mathrm{Fin}) \models \forall z_1\dots z_m\ \forall y_1\dots y_m(\chi(y_1,\dots,y_m) \wedge \bigwedge_{1\le j\le m} y_j \le z_j \rightarrow \chi(z_1,\dots,z_m)).$$

Proof. The proof is by induction on φ. For atomic φ, $\varphi = \neg\varphi'$, $\varphi = \varphi_1 \wedge \varphi_2$ or $\varphi = \exists x\varphi'$ one can argue as in the proof of 6.3.2 in [3]. The formulas $\psi_1,\dots,\psi_m$ obtained there are L_t-formulas having the additional property, that any set variable occuring positively (negatively) in a ψ_j occurs positively (negatively) in φ.

Now assume that φ is $\exists Y \ni t\varphi'$, $\varphi' = \varphi'(\bar{x},\bar{X}^+,\bar{Y}^-,Y^-)$ and that φ' is determined by $(\chi';\psi_1,\dots,\psi_m)$. Then $\psi_1,\dots,\psi_m$ are negative in Y. Let $l = 2^m$ and let $s_1,\dots,s_l$ be a listing of all subsets of $\{1,\dots,m\}$ with $s_j = \{j\}$ for $1 \le j \le m$. - For $1 \le h \le l$, let

$$\rho_h = \exists Y \ni t \bigwedge_{j\in s_h} \psi_j ,$$

and for $1 \le j \le m$, let

$$\vartheta_j = \psi_j \frac{T}{Y} ,$$

(where $\psi_j \frac{T}{Y}$ is obtained from ψ_j substituting any atomic part of the form $t' \in Y$ by $t' = t'$).

Let $\chi = \chi(y_1,\dots,y_l,v_1,\dots,v_m)$ be

$$\chi = \exists z_1 \ldots \exists z_l (\bigwedge_{1 \leq h \leq l} z_h \leq y_h \wedge \bigwedge_{s_j \cup s_k = s_n} z_j \cap z_k = z_h$$

$$\wedge \ \chi'(z_1,\ldots,z_m) \wedge \bigwedge_{1 \leq j \leq m} \mathrm{Fin}(z_j - v_j)).$$

The monotonicity of χ' implies that of χ. We show that $\exists Y \ni t\varphi'$ is determined by $(\chi;\varrho_1,\ldots,\varrho_l,\vartheta_1,\ldots,\vartheta_m)$. - Let us first suppose that $\prod_I (\mathfrak{A}_i,\sigma_i) \models \exists Y \ni t\ \varphi'[\bar{a},\bar{U},\bar{V}]$, say $\prod_I (\mathfrak{A}_i,\sigma_i) \models \varphi'[\bar{a},\bar{U},\bar{V},V]$ where $V \in \mathring{\prod_I} \sigma_i$.

For $1 \leq h \leq l$, let

(1) $M_h = S(\varrho_h) = \{i \in I | (\mathfrak{A}_i,\sigma_i) \models \exists Y \ni t \bigwedge_{j \in s_h} \psi_j[\bar{a}(i),\bar{U}(i),\bar{V}(i)]\}$,

and for $1 \leq j \leq m$, let

(2) $N_j = S(\vartheta_j) = \{i \in I | (\mathfrak{A}_i,\sigma_i) \models \psi_j{}^{T}_{Y}[\bar{a}(i),\bar{U}(i),\bar{V}(i)]\}$.

We see that $(\mathfrak{P}(I),\mathrm{Fin}) \models \chi[M_1,\ldots,M_l,N_1,\ldots,N_m]$, taking as $z_1,\ldots,z_l$ the sets

$$Z_h = \{i \in I | (\mathfrak{A}_i,\sigma_i) \models \bigwedge_{j \in s_h} \psi_j[\bar{a}(i),\bar{U}(i),\bar{V}(i),V(i)]\}.$$

Conversely, suppose

(3) $(\mathfrak{P}(I),\mathrm{Fin}) \models \chi[M_1,\ldots,M_l,N_1,\ldots,N_m]$

Where $M_1,\ldots,N_m$ are defined by (1) and (2). We show that $\prod_I(\mathfrak{A}_i,\sigma_i) \models \exists Y \ni t\ \varphi'[\bar{a},\bar{U},\bar{V}]$. By (3) we find Z_h, $1 \leq h \leq l$, satisfying the "matrix" of χ. In particular

(4) $(\mathfrak{P}(I),\mathrm{Fin}) = \chi'[Z_1,\ldots,Z_m]$.

For $i \notin \bigcup_{j=1}^{m} (Z_j - N_j)$, let $V_i = A_i$. For $i \in \bigcup_{j=1}^{m} (Z_j - N_j)$, let $s = \{j | i \in Z_j,\ 1 \leq j \leq m\}$. Choose h such that $s = s_h$. Since $Z_h \subset M_h$ there is $V_i \in \sigma_i$ with

$$t^{(\mathfrak{A}_i,\sigma_i)}[\bar{a}(i)] \in V_i \quad \text{and} \quad (\mathfrak{A}_i,\sigma_i) \models \bigwedge_{j \in s_h} \psi_j[\bar{a}(i),\bar{U}(i),\bar{V}(i),V_i].$$

Let V be $\prod_{i \in I} V_i$. Then $t^{\Pi(\mathfrak{A}_i,\sigma_i)}[\bar{a}] \in V$ and since $\bigcup_{j=1}^{m} (Z_j - N_j)$ is finite, $V \in \overset{o}{\underset{I}{\Pi}} \sigma_i$.

For $1 \leq j \leq m$ define H_j by

$$H_j = \{i \mid (\mathfrak{A}_i,\sigma_i) \models \psi_j[\bar{a}(i),\bar{U}(i),\bar{V}(i),V(i)].$$

Now one easily verifies that $Z_j \subset H_j$. Hence, by (4) and monotonicity of χ'

$$(\mathfrak{P}(I),Fin) \models \chi'[H_1,\ldots,H_m].$$

Therefore,

$$\underset{I}{\Pi}(\mathfrak{A}_i,\sigma_i) \models \varphi'[\bar{a},\bar{U},\bar{V},V], \qquad \text{hence} \quad \underset{I}{\Pi}(\mathfrak{A}_i,\sigma_i) \models \exists Y \ni t\ \varphi'[\bar{a},\bar{U},\bar{V}].$$

Given a class $\mathfrak{K}$ of topological structures, denote by $Th_t(\mathfrak{K})$ the <u>L_t-theory of $\mathfrak{K}$</u>, i.e.

$$Th_t(\mathfrak{K}) = \{\varphi \mid \varphi \in L_t, (\mathfrak{A},\sigma) \models \varphi \text{ for all } (\mathfrak{A},\sigma) \in \mathfrak{K}\}.$$

In case $\mathfrak{K} = \{(\mathfrak{A},\sigma)\}$, we write $Th_t(\mathfrak{A},\sigma)$.

Using 7.1 and 7.2 of [5] we obtain from the preceding theorem:

6.4 <u>Corollary</u>. (1) Suppose that $Th_t(\mathfrak{K})$ is decidable. Let $\mathfrak{K}'$ be the class of all products of members of $\mathfrak{K}$. Then $Th_t(\mathfrak{K}')$ is decidable.

(2) The class of L_t-sentences preserved under finite, the class of L_t-sentences preserved under infinite, and the class of L_t-sentences preserved under arbitrary products are recursively enumerable.

(Since for any $\varphi \in L_t$ the set $\{\psi \mid \models_t \varphi \rightarrow \neg \psi\}$ is recursively enumerable, (2) is obtained from 7.1 in [5].) Note that $(\exists x\, Px \wedge \exists x\, Qx \wedge \forall x(Px \leftrightarrow \neg Qx) \vee \neg \text{disc})$ is an L_t-sentence preserved under infinite but not under all finite products.

6.5 <u>Corollary</u>. $Th_t((\mathfrak{A},\sigma)^I)$ is decidable for any finite structure $(\mathfrak{A},\sigma)$.

In particular, let 2 be $\{0,1\}$ and take the discrete topology σ on 2. Then $Th_t((2,\sigma)^I)$ is decidable.

One can derive the corresponding results for "weak direct products" (direct sums): Assume that L contains just one constant, say c. We restrict our attention to structures satisfying $fc\ldots c = c$ for any $f \in L$. The <u>direct sum</u> $\bigoplus_I (\mathfrak{A}_i, \sigma_i)$ of topological structures $(\mathfrak{A}_i, \sigma_i)$ is defined to be the substructure of $(\prod_I \mathfrak{A}_i, \tilde{\prod}_I \sigma_i)$ (where $\tilde{\prod}_I \sigma_i$ denotes the "box"-topology) with universe

$$\{f \mid f \in \prod A_i, f(i) = c^{\mathfrak{A}_i} \text{ almost everywhere}\}.$$

$\bigoplus_I (\mathfrak{A}_i, \sigma_i)$ is a relativized generalized product relative to $(\mathfrak{P}(I), Fin)$. It is possible to prove 6.3 for direct sums, i.e. any $\varphi \in L_t$ is determined in a direct sum by some $(\chi, \psi_1, \ldots, \psi_m)$ with $\psi_i \in L_t$. Since the correspondence $\varphi \mapsto (\chi, \psi_1, \ldots, \psi_m)$ is effective, 6.4 is also true for direct sums.

We leave it to the reader to formulate and prove a general L_t-Feferman-Vaught-theorem, which contains the results on products, sums and direct sums as special cases.

"Syntactic characterizations" of the sentences preserved by the algebraic operations of this section and by related operations are not known. In particular this is true for the sentences preserved by products and substructures or for the sentences preserved by direct factors; in both cases, easy characterizations for $L_{\omega\omega}$ are known.-

By the way we remark that there is no characterization of the L_t-sentences preserved by the intersection (resp. union) of topologies, i.e. of those $\varphi \in L_t$ such that

$$(\mathfrak{A}, \sigma_1) \models \varphi \text{ and } (\mathfrak{A}, \sigma_2) \models \varphi \text{ imply } (\mathfrak{A}, \sigma_1 \cap \sigma_2) \models \varphi$$

(resp. $(\mathfrak{A}, \sigma_1) \models \varphi$ and $(\mathfrak{A}, \sigma_2) \models \varphi$ imply $(\mathfrak{A}, \sigma_1 \uplus \sigma_2) \models \varphi$ where $\sigma_1 \uplus \sigma_2$ is the coarsest topology containing $\sigma_1 \cup \sigma_2$).

6.6 <u>Exercise</u>. Show that - for recursive L - the class of L_t-sentences preserved under the intersection (resp. union) of topologies is recursively enumerable.

The result of $L_{\omega\omega}$ that a convex elementary class is closed under the union of structures does not generalize to L_t, as it is shown by the next exercise.

6.7 Exercise. For $L = \{c_o, c_1\}$ let $\varphi_o = \exists X \ni c_o\ \exists Y \ni c_1\ \forall z(\neg z \in X \vee \neg z \in Y)$. Show: (a) Given $(\mathfrak{B},\sigma) \models \varphi_o$ and substructures $(\mathfrak{A}_1,\sigma_1),(\mathfrak{A}_2,\sigma_2)$ and $(\mathfrak{A}_3,\sigma_3)$ of $(\mathfrak{B},\sigma)$; if $(\mathfrak{A}_1,\sigma_1) \models \varphi_o$, $(\mathfrak{A}_2,\sigma_2) \models \varphi_o$ and $A_1 \cap A_2 = A_3$, then $(\mathfrak{A}_3,\sigma_3) \models \varphi_o$.

(b) There is a sequence $(\mathfrak{A}_n,\sigma_n)$ of models of φ_o with $(\mathfrak{A}_n,\sigma_n) \subset (\mathfrak{A}_{n+1},\sigma_{n+1})$ such that the union of the $(\mathfrak{A}_n,\sigma_n)$ (i.e. the injective limit in the category of topological spaces) is not a model of φ_o.

§ 7 Definability

First we show that some theorems on the explicit definability of relations generalize from $L_{\omega\omega}$ to L_t. After that we prove some results on the explicit definability of topologies. - For simplicity, we assume in this section that all similarity types are denumerable.

7.1 Beth's theorem. Let T be an L_t'-theory, $L \subset L'$ and $R \in L' - L$. The following are equivalent:

(i) If $(\mathfrak{A},\sigma) \models T$, $(\mathfrak{B},\tau) \models T$ and $(\mathfrak{A}\restriction L,\sigma) \models (\mathfrak{B}\restriction L,\tau)$ then $R^{\mathfrak{A}} = R^{\mathfrak{B}}$.

(ii) There is a $\varphi(\bar{x}) \in L_t$ such that $T \models_t \forall x(R\bar{x} \leftrightarrow \varphi(\bar{x}))$.

Since the proof - using the interpolation theorem - is the same as for $L_{\omega\omega}$, we omit it.

We call an L_t-theory T complete, if for any L_t-sentece φ either $T \models_t \varphi$ or $T \models_t \neg\varphi$.

7.2 Svenonius's theorem. Let T be an L_t'-theory, $L \subset L'$ and $R \in L' - L$. The following are equivalent:

(i) If $(\mathfrak{A},\sigma) \models T$ and π is an automorphism of $(\mathfrak{A}\restriction L,\sigma)$, then π maps $R^{\mathfrak{A}}$ onto itself (i.e. $R^{\mathfrak{A}} = \{\pi(\bar{a}) \mid \bar{a} \in A, R^{\mathfrak{A}}\bar{a}\}$).

(ii) There are a finite number of formulas $\varphi_1(\bar{x}),\ldots,\varphi_n(\bar{x}) \in L_t$ such that

$$T \models_t \bigvee_{i=1}^{n} \forall\bar{x}(R\bar{x} \leftrightarrow \varphi_i(\bar{x})).$$

Proof. (ii) $\Rightarrow$ (i). This is trivial.

(i) $\Rightarrow$ (ii). Suppose, by contradiction, that (ii) does not hold.

Then

$$T_o := T \cup \{\neg \forall \bar{x}(R\bar{x} \leftrightarrow \varphi(\bar{x})) \mid \varphi = \varphi(\bar{x}) \in L_t\}$$

has a topological model. Let $T' \subset L_t'$ be any complete extension of T_o. Since for T' part (ii) of Beth's theorem does not hold, there are models $(\mathfrak{A},\sigma)$ and $(\mathfrak{B},\tau)$ of T' with $(\mathfrak{A}\restriction L,\sigma) = (\mathfrak{B}\restriction L,\tau)$ and $R^{\mathfrak{A}} \neq R^{\mathfrak{B}}$. Choose a denumerable recursively saturated (weak) model $((\mathfrak{C},(k^*)^{\mathfrak{C}}_{k \in L'-L}),\sigma_1)$ of $Th_t((\mathfrak{A},(k^{\mathfrak{B}})_{k \in L'-L}),\sigma))$. Since T' is complete, $(\mathfrak{A},\sigma) \equiv^t (\mathfrak{B},\tau)$ and therefore by 4.7, $((\mathfrak{C}\restriction L,R^{\mathfrak{C}}),\tilde{\sigma}_1)$ and $((\mathfrak{C}\restriction L,R^{*\mathfrak{C}}),\tilde{\sigma}_1)$ are homeomorphic. Hence there is an automorphism of $(\mathfrak{C}\restriction L,\tilde{\sigma}_1)$ mapping $R^{\mathfrak{C}}$ onto $R^{*\mathfrak{C}}$. But $R^{\mathfrak{C}} \neq R^{*\mathfrak{C}}$, contradicting (i).

In § 8 we show that Chang-Makkai-theorem does not generalize to L_t.

Now we start the study of problems concerning the definability of topologies. Suppose that T is an L_t-theory which <u>defines the topology implicitly</u>, i.e.

$$(\mathfrak{A},\sigma) \models T \text{ and } (\mathfrak{A},\tau) \models T \text{ imply } \sigma = \tau.$$

An application of the interpolation theorem then shows that each L_t-sentence is equivalent in T to an $L_{\omega\omega}$-sentence : namely let L' be a similarity type appropriate for structures of the form $(\mathfrak{B},\sigma_1,\sigma_2)$, where $\mathfrak{B}$ is an L-structure. Given $\psi \in L_t$ denote by ψ^1 resp. ψ^2 an L_t'-sentence such that

$$(\mathfrak{B},\sigma_1,\sigma_2) \models \psi^i \quad \text{iff} \quad (\mathfrak{B},\sigma_i) \models \psi.$$

Since T defines the topology implicitly, we have for any $\psi \in L_t$

$$T^1 \cup T^2 \models_t \varphi^1 \to \varphi^2,$$

where $T^i = \{\psi^i \mid \psi \in T\}$. Therefore for some finite $T_o \subset T$,

$$\models_t \wedge T_o^{\,1} \wedge \varphi^1 \to (\wedge T_o^{\,2} \to \varphi^2).$$

Using the interpolation for L_t' (see 5.3) we obtain a $\psi \in L_{\omega\omega}$ with

$$\models_t \wedge T_o^{\,1} \wedge \varphi^1 \to \psi \quad \text{and} \quad \models_t \psi \to (\wedge T_o^{\,2} \to \varphi^2).$$

Hence $T \models_t \varphi \leftrightarrow \psi$.

We get a "uniform" translation from L_t to $L_{\omega\omega}$ by the next theorem stating that T defines the topology "explicitly" (if T defines it implicitly).

Given an $L_{\omega\omega}$-formula $\varphi(x,\bar{y})$, a structure $\mathfrak{A}$ and $\bar{a} \in A$, we denote by $\varphi^{\mathfrak{A}}(\cdot,\bar{a})$ the set

$$\varphi^{\mathfrak{A}}(\cdot,\bar{a}) = \{b \mid b \in A, \mathfrak{A} \models \varphi[b,\bar{a}]\},$$

and by $\varphi^{\mathfrak{A}}$ the collection of sets

$$\varphi^{\mathfrak{A}} = \{\varphi^{\mathfrak{A}}(\cdot,\bar{a}) \mid \bar{a} \in A\}.$$

7.3 <u>Definition</u>. Let T be an L_t-theory and $\varphi(x,\bar{y})$ an $L_{\omega\omega}$-formula. φ <u>explicitly defines the topology</u> in T, if for any model $(\mathfrak{A},\sigma)$ of T we have $\sigma = \widetilde{\varphi^{\mathfrak{A}}}$, i.e. if $T \models_t \text{basis}(\varphi)$ where

$$\text{basis}(\varphi) = \forall x\, \forall \bar{y}(\varphi(x,\bar{y}) \to \exists X \ni x(\forall z(z \in X \to \varphi(z,\bar{y})))$$
$$\wedge\ \forall x\, \forall X \ni x\ \exists \bar{y}(\varphi(x,\bar{y}\ )\ \wedge\ \forall z(\varphi(z,\bar{y}) \to z \in X))).$$

7.4 <u>Example</u>. Let $L = \{<\}$ and let T_o be the theory of linearly ordered sets without endpoints, carrying the order topology. Thus T_o contains besides $L_{\omega\omega}$-axioms the L_t-sentences

$$\forall x\ \forall y\ \forall z\ (y < x < z \to \exists X \ni x\ \forall u\ (u \in X \to y < u < z))$$

$$\forall x\ \forall X \ni x\ \exists y\ \exists z\ (y < x < z \wedge \forall u\ (y < u < z \to u \in X)).$$

The $L_{\omega\omega}$-formula $\varphi(x,y_1,y_2) = y_1 < x < y_2$ explicitly defines the topology in T_o.

If $\varphi(x,\bar{y})$ explicitly defines the topology in T, then to any L_t-sentence ψ we obtain an equivalent $L_{\omega\omega}$-sentence ψ^{φ} eliminating set quantifiers as indicated by

$$\exists X \ni t\ldots t' \in X \text{ ---}\mapsto \exists \bar{y}(\varphi(t,\bar{y}) \wedge \ldots \varphi(t',\bar{y}) \text{ --- })$$
$$\forall X \ni t\ldots t' \in X \text{ ---}\mapsto \forall \bar{y}(\varphi(t,\bar{y}) \to \ldots \varphi(t',\bar{y}) \text{ --- }).$$

Hence

7.5 <u>Theorem</u>. Let T be an L_t-theory and suppose $\varphi(x,\bar{y})$ explicitly defines the topology in T. Let T^{φ} be the $L_{\omega\omega}$-theory $\{\psi^{\varphi} \mid \psi \in T\}$. Then:

(a) T and $T^{\varphi} \cup \{\text{basis}(\varphi)\}$ have the same models.

(b) For any L-structure $\mathfrak{A}$,

$$\mathfrak{A} \models T^{\varphi} \cup \{(\underline{\text{bas}})^{\varphi}\} \quad \text{iff} \quad (\mathfrak{A},\sigma) \models T \text{ for some topology } \sigma \text{ on } A.$$

We now prove

7.6 Ziegler's definability theorem. Given an L_t-theory T the following are equivalent:

(i) T defines the topology implicitly.

(ii) The topology is explicitly definable in T.

Proof. The proof of (i) from (ii) is easy. - Now, assume (i). Choose a new unary relation symbol P and a new constant d. First we show:

(1) There is a sentence $\chi \in (L \cup \{P,d\})_{\omega\omega}$, $\chi = \chi(P,d)$ such that for any model $(\mathfrak{A},\sigma)$ of T, any $B \subset A$ and any $a \in A$, we have

$$((\mathfrak{A},B,a),\sigma) \models \chi \text{ iff } B \text{ is a (not necessarily open) neighborhood of } a.$$

To prove (1) let ψ be an $(L \cup \{P,d\})_t$-sentence expressing

"P is a neighborhood of d".

Since T defines the topology implicitly- also as $(L \cup \{P,d\})_t$-theory -, there is (compare the remarks after 7.2) a $\chi \in (L \cup \{P,d\})_{\omega\omega}$ such that $T \models_t \psi \leftrightarrow \chi$.

Thus, for any $L_{\omega\omega}$-formula $\Theta(z,\bar{w})$

"if $\Theta(\cdot,\bar{w})$ is a neighborhood of y, then $\Theta(\cdot,\bar{w}) \not\subset Y$"

is expressed by the $L_{\omega\omega}$-formula

$$\chi(\Theta(\cdot,\bar{w}),y) \to \neg \forall z(\Theta(z,\bar{w}) \to z \in Y),$$

where $\chi(\Theta(\cdot,\bar{w}),y)$ is obtained from $\chi(P,d)$ substituting y for d and replacing each formula Pt by $\Theta(t,\bar{w})$.

Now suppose that (ii) does not hold. Then the set

$$\Phi := T \cup \{y \in Y\} \cup \{\forall\bar{w}(\chi(\Theta(\cdot,\bar{w}),y) \to \neg \forall z(\Theta(z,\bar{w}) \to z \in Y)) \mid \Theta(z,\bar{w}) \in L_{\omega\omega}\}$$

has a topological model, in which the interpretation of Y is an open set. Otherwise for some $\Theta_1,\dots,\Theta_n \in L_{\omega\omega}$, $\Theta_i = \Theta_i(z,\bar{w})$,

$$T \cup \{y \in Y\} \models_t \bigvee_{i=1}^{n} \exists\bar{w}(\chi(\Theta_i(\cdot,\bar{w}),y) \wedge \forall z(\Theta_i(z,\bar{w}) \to z \in Y))$$

and therefore

$$T \models_t \forall y\ \forall Y \ni y \bigvee_{i=1}^{n} \exists\bar{w}(\chi(\Theta_i(\cdot,\bar{w}),y) \wedge \forall z(\Theta_i(z,\bar{w}) \to z \in Y)).$$

Put

$$\Theta(z,\bar{w},v_o,\dots,v_n) := \bigvee_{i=1}^{n} (v_o = v_i \wedge \Theta_i(z,\bar{w})).$$

Then in any model of T (with more than one element) each open set contains with each of its points a neighborhood of this point, which is a "Θ-set". Let $\Theta'(z,\hat{w})$ be the formula (where $\hat{w}$ is $\bar{w},v_o,\dots,v_n$)

$$\Theta'(z,\hat{w}) := \chi(\Theta(\cdot,\hat{w}),z).$$

Then $\Theta'(z,\hat{w})$ explicitly defines the topology in T.

Thus, assume Φ is satisfiable. Let L' be a similarity type appropriate for structures of the form $(\mathfrak{A},\sigma_1,\sigma_2)$, where $\mathfrak{A}$ is an L-structure. Let X,X',... denote set variables for the first topology, Y,Y',... set variables for the second topology. Let T^1 (resp. T^2) be a set of L'_t-sentences containing only set variables of the first (resp. second) sort such that

$$(\mathfrak{A},\sigma_1,\sigma_2) \models T^i \quad \text{iff} \quad (\mathfrak{A},\sigma_i) \models T \ .$$

Choose new constants $c_o,c_1,\dots,$ set constants $U_o,U_1,\dots$ (for the first sort), and $V_o,V_1,\dots$ (for the second sort). The set constants will be interpreted in models by open sets of the corresponding topology.

Let T' be the set

$$T' = T^1 \cup T^2 \cup \{c_o \in V_o\} \cup \{\forall\bar{w}(\chi(\Theta(\cdot,\bar{w}),c_o) \to \neg\forall z(\Theta(z,\bar{w}) \to z \in V_o)) \mid \Theta(z,\bar{w}) \in L_{\omega\omega}\}.$$

Since Φ has a model, so does T'. As usual one can construct step by step a sequence

$$T' \subset T'' \subset T''' \subset \dots$$

of consistent sets such that $T^\infty := \bigcup_{n \geq 1} T^{(n)}$ is a Hinitkka set (see 1.4), where in each formula of T^∞ all set variables and set constants are of the same sort. We arrange the construction in such a way that whenever we add for some constant U (= U_i) the formula $c_o \in U$ we also add for some new constant c (= c_m) the formulas

(2) $$c \in U, \neg c \in V_o.$$

This is possible: namely assume that

(3) $$\{c_o \in U\} \subset T^{(n)}$$

has a model. We show that $T^{(n)} \cup \{c \in U\} \cup \{\neg c \in V_o\}$ has a model.

Otherwise

$$T^{(n)} \models_t c \in U \to c \in V_o.$$

Write $T^{(n)}$ as the union of $T_1^{(n)}$ and $T_2^{(n)}$, where $T_i^{(n)}$ contains only set variables and set constants of sort i. Then

$$\models_t (\wedge T_1^{(n)} \wedge c \in U) \to (\wedge T_2^{(n)} \to c \in V_o)$$

(using the compactness theorem we can replace infinite conjunctions by finite conjunctions). By the interpolation theorem (see 5.3), we obtain $\varphi(c,\bar{c}) \in (L \cup \{c,\bar{c}\})_{\omega\omega}$ such that

$$T_1^{(n)} \models_t c \in U \to \varphi(c,\bar{c}) \quad \text{and} \quad T_2^{(n)} \models_t \varphi(c,\bar{c}) \to c \in V_o.$$

Since c does not occur in $T_1^{(n)} \cup T_2^{(n)}$, we have

(4) $$T^{(n)} \models_t \forall z(z \in U \to \varphi(z,\bar{c}))$$

(5) $$T^{(n)} \models_t \forall z(\varphi(z,\bar{c}) \to z \in V_o).$$

From (4), (3) and (1) we obtain

(6) $$T^{(n)} \models_t \chi(\varphi(\cdot,\bar{c}),c_o).$$

Since $\{\forall\bar{x}(\chi(\varphi(\cdot,\bar{w}),c_o) \to \neg \forall z(\varphi(z,\bar{w}) \to z \in V_o))\} \subset T^{(n)}$, we get from (5)

(7) $$T^{(n)} \models_t \neg \chi(\varphi(\cdot,\bar{c}),c_o).$$

By (6) and (7), $T^{(n)}$ has no model, a contradiction

Now, let $(\mathfrak{B},\tau_1,\tau_2)$ be the (weak) term model associated with the Hintikka set T^{∞}. Then $(\mathfrak{B},\tilde{\tau}_1)$ and $(\mathfrak{B},\tilde{\tau}_2)$ are models of T. But by construction (see (2) and (3)) of T^{∞}, we know that V_o is a $\tilde{\tau}_2$-open neighborhood of $c_o^{\mathfrak{B}}$ which contains no $\tilde{\tau}_1$-open neighborhood of $c_o^{\mathfrak{B}}$. Therefore $\tilde{\tau}_1 \neq \tilde{\tau}_2$ and hence, T does not define the topology implicitly.

7.7 Remark. Adding in the preceding proof to L' a disjoint copy of $L - L_o$, will obtain:

Let T be an L_t-theory and $L_o \subset L$. The following are equivalent:

(i) If $(\mathfrak{A},\sigma) \models T$, $(\mathfrak{B},\tau) \models T$ and $\mathfrak{A}\restriction L_o = \mathfrak{B}\restriction L_o$, then $\sigma = \tau$.

(ii) There is an $(L_o)_{\omega\omega}$-formula, which defines explicitly the topology in T.

It is easy to derive Svenonius theorem for topologies from 7.6 in the same way as 7.2 was obtained from 7.1, or by a proof similar to that of the preceding theorem where we start with a complete T' and get finally a homeomorphism between both topologies as the union of intended partial homeomorphisms $p_o \subset p_1 \subset \dots$. Here p_n is fixed after defining $T^{(n)}$. It is useful to choose $T^{(n)}$ complete with respect to the language containing the finitely many constants used so far.

In L_t a Chang-Makkai-theorem for topologies holds: Given a formula $\varphi(x,\bar{y},\bar{w})$ let basis(φ) be the formula

$$\begin{aligned} \text{basis}(\varphi) = {} & \forall x\, \forall \bar{y}(\varphi(x,\bar{y},\bar{w}) \rightarrow \exists X \ni x\, \forall z(z \in X \rightarrow \varphi(z,\bar{y},\bar{w})) \\ & \wedge \forall x\, \forall X \ni x\, \exists \bar{y}(\varphi(x,\bar{y},\bar{w}) \wedge \forall z(\varphi(z,\bar{y},\bar{w}) \rightarrow z \in X)). \end{aligned}$$

Thus the free variables of basis(φ) are among $\bar{w}$.

7.8 <u>Theorem</u>. Given an L_t-theory T, the following are equivalent:

(i) For every denumerable $\mathfrak{A}$,

$$|\{\sigma \mid \sigma \text{ topology on } A, (\mathfrak{A},\sigma) \models T\}| < 2^{\aleph_o}.$$

(ii) For every model $(\mathfrak{A},\sigma)$ of T with denumerable A,

$$|\{\tau \mid \tau \text{ topology on } A, (\mathfrak{A},\sigma) \simeq^t (\mathfrak{A},\tau)\}| < 2^{\aleph_o}.$$

(iii) There are a finite number of $L_{\omega\omega}$-formulas $\varphi_1(x,\bar{y},\bar{w}),\dots,\varphi_r(x,\bar{y},\bar{w})$ such that

$$T \models_t \bigvee_{i=1}^{r} \exists \bar{w}\, \text{basis}(\varphi_i).$$

<u>Sketch of proof</u>. It is easy to prove the implications (iii) ⇒ (i) and (i) ⇒ (ii).

(ii) ⇒ (iii). Suppose by contradiction, that (iii) does not hold.

Then

$$T' := T \cup \{\forall \bar{w} \neg \text{basis}(\varphi) \mid \varphi(x,\bar{y},\bar{w}) \in L_{\omega\omega}\}$$

has a model. As usual (compare [2],[14]) by a tree argument we construct a Hinitkka set leading to $2^{\aleph_o}$ topologies. As tree one can choose $(2^{\underline{\omega}},\subset)$ (the set of finite sequences of 0's and 1's ordered by inclusion). For any branch

α in $(2^{\omega},c)$, i.e. for $\alpha \in 2^{\omega}$, we introduce set variables $X^{\alpha},Y^{\alpha},\ldots$. For $\varphi \in L_t$ let φ^{α} be obtained from φ by replacing the set variables by $X^{\alpha},Y^{\alpha},\ldots$. We have to distinguish two cases:

Case_1: There is an $L_{\omega\omega}$-formula $\varphi(x,\bar{y})$ such that

$$T^* := T' \cup \{\neg \exists \bar{v}\, \forall \bar{y}(\text{"}\varphi(\cdot,\bar{y}) \text{ open"} \leftrightarrow \psi(\bar{y},\bar{v})) \mid \psi(\bar{y},\bar{v}) \in L_{\omega\omega}\}$$

has a model. Here "$\varphi(\cdot,\bar{y})$ open" stands for

$$\forall x(\varphi(x,\bar{y}) \rightarrow \exists X \ni x\ \forall z(z \in X \rightarrow \varphi(z,\bar{y}))).$$

Then the construction is arranged in such a way that for any distinct branches $\alpha,\beta \in 2^{\omega}$ there are constants $\bar{c}$ such that the formulas

$$\text{"}\varphi(\cdot,\bar{c}) \text{ open"}^{\alpha} \quad \text{and} \quad \neg\, \text{"}\varphi(\cdot,\bar{c}) \text{ open"}^{\beta}$$

belong to the Hintikka set (use the interpolation theorem).

Case_2. For every $\varphi(x,\bar{y}) \in L_{\omega\omega}$ there is an $L_{\omega\omega}$-formula, which we denote by $\psi_{\varphi}(\bar{y},\bar{v})$, such that

$$T' \models_t \exists \bar{v}\, \forall \bar{y}(\text{"}\varphi(.,\bar{y}) \text{ open"} \leftrightarrow \psi_{\varphi}(\bar{y},\bar{v})).$$

Similarly as in the proof of the consistency of Φ in 7.6 one can show that for new constans $\bar{c}$

$$\Phi^* := T' \cup \{x \in X \wedge \forall \bar{y}(\text{"}\varphi(\cdot,\bar{y}) \text{ open"} \leftrightarrow \psi_{\varphi}(\bar{y},\bar{c}))\}$$
$$\cup \{\forall \bar{y}(\varphi(x,\bar{y}) \wedge \psi_{\varphi}(\bar{y},\bar{c}) \rightarrow \neg \forall z(\varphi(z,\bar{y}) \rightarrow z \in X)) \mid \varphi(x,\bar{y}) \in L_{\omega\omega}\}$$

has a model. When constructing the Hintikka set we handle any two different branches of the tree similarly as indicated after 7.6 for the analogue of Svenonius' theorem.

It is not difficult for $L_{\omega\omega}$ to derive from Chang-Makkai's theorem the definability theorems of Beth, Svenonius, Kueker,... . We demonstrate this method for L_t, deriving from 7.8 Svenonius' theorem for topologies and then sketch a proof of Kuekers theorem for topologies.

We introduce some notation. We write $\mathfrak{A} \simeq_n \mathfrak{B}$ if $\mathfrak{A}$ and $\mathfrak{B}$ are n-isomorphic (in the algebraic sense). For $\bar{a} \in A$, $\varphi^n_{\bar{a}}$ denotes the n-isomorphism type of $\bar{a}$ in $\mathfrak{A}$ (cf. [6]). - For $\varphi(x,\bar{y},\bar{w}) \in L_{\omega\omega}$ and any $\bar{a} \in A$ let $\varphi^{\mathfrak{A}}(\cdot,-,\bar{a})$ be the set of all subsets $\varphi^{\mathfrak{A}}(\cdot,\bar{b},\bar{a})$ of A, where $\bar{b} \in A$.

7.9 Theorem. Let T be an L_t-theory. Then the following are equivalent:

(i) For every model $(\mathfrak{A},\sigma)$ of T,

$$|\{\tau \mid \tau \text{ topology on } A, (\mathfrak{A},\sigma) \simeq^t (\mathfrak{A},\tau)\}| = 1 .$$

(ii) There are finitely many $L_{\omega\omega}$-formulas $\psi_1(x,\bar{y}),\ldots,\psi_r(x,\bar{y}) \in L_{\omega\omega}$ such that

$$T \models_t \bigvee_{i=1}^{r} \text{basis}(\psi_i)$$

Proof. Suppose that (i) holds. Using a compactness argument we may assume that L is finite. By 7.8 there are $\varphi_1(x,\bar{y},\bar{w}),\ldots,\varphi_r(x,\bar{y},\bar{w})$ such that

(1) $$T \models_t \bigvee_{i=1}^{r} \exists\bar{w}\ \text{basis}(\varphi_i).$$

We show that

for each $i = 1,\ldots,r$ there is an n_i such that for any $(\mathfrak{A},\sigma) \models T$,

(2) any $\bar{a},\bar{b} \in A$, we have

$$\text{if } \sigma = \widetilde{\varphi_i^{\mathfrak{A}}(\cdot,-,\bar{a})} \text{ and } (\mathfrak{A},\bar{a}) \simeq_{n_i} (\mathfrak{A},\bar{b}), \text{ then}$$

$$\widetilde{\varphi_i^{\mathfrak{A}}(\cdot,-,\bar{a})} = \widetilde{\varphi_i^{\mathfrak{A}}(\cdot,-,\bar{b})} .$$

Otherwise, applying the compactness theorem and the Löwenheim-Skolem theorem as in the convergence lemma, we obtain a model $(\mathfrak{A},\sigma)$ of T and $\bar{a},\bar{b} \in A$ such that

$$\sigma = \widetilde{\varphi_i^{\mathfrak{A}}(\cdot,-,\bar{a})},\ (\mathfrak{A},\bar{a}) \simeq (\mathfrak{A},\bar{b}),\ \text{say } \pi\colon (\mathfrak{A},\bar{a}) \simeq (\mathfrak{A},\bar{b}) \text{ and}$$

$$\widetilde{\varphi_i^{\mathfrak{A}}(\cdot,-,\bar{a})} \neq \widetilde{\varphi_i^{\mathfrak{A}}(\cdot,-,\bar{b})} .$$

But then, $\pi\colon (\mathfrak{A},\sigma) \simeq^t (\mathfrak{A},\widetilde{\varphi_i^{\mathfrak{A}}(\cdot,-,\bar{b})})$ and $\sigma \neq \widetilde{\varphi_i^{\mathfrak{A}}(\cdot,-,\bar{b})}$, contrary to (i).

By (1) and (2) we obtain

$$T \models_t \bigvee_{\substack{i = 1,\ldots r \\ \mathfrak{A} \text{ and } \bar{a} \in A \\ \text{arbitrary}}} \text{basis}(\exists\bar{w}(\varphi_i(\cdot,-,\bar{w}) \wedge \varphi_{\bar{a}}^{n_i}(\bar{w}))).$$

7.10 Theorem. Let T be an L_t-theory and let $n \geq 1$. Then the following are equivalent:

(i) For every structure $\mathfrak{A}$,

$$|\{\sigma | \sigma \text{ topology on } A, (\mathfrak{A},\sigma) \models T\}| \leq n.$$

(ii) There are $L_{\omega\omega}$-formulas $\chi(v_1,\ldots,v_k)$ and $\Theta_i(x,\bar{y},v_1,\ldots,v_k)$, $1 \leq i \leq n$ such that $T \models_t \exists v_1\ldots v_k\ \chi(v_1,\ldots,v_k)$ and

$$T \models_t \forall v_1\ldots v_k(\chi \rightarrow \bigvee_{1 \leq i \leq n} \text{basis}(\Theta_i)).$$

<u>Proof</u>. The implication (ii) ⇒ (i) is easy to verify. (i) ⇒ (ii). Assume that (i) holds. By an easy compactness argument, we may assume that L is finite and that $T = \{\varphi\}$. 7.8 tells us that there are $\varphi_1(x,\bar{y},\bar{w}),\ldots,\varphi_r(x,\bar{y},\bar{w})$ such that

(1) $$\varphi \models_t \bigvee_{1 \leq i \leq r} \exists\bar{w}\ \text{basis}(\varphi_i)\ .$$

One can prove that

for each $i = 1,\ldots,r$ there is an n_i such that for any $(\mathfrak{A},\sigma) \models \varphi$

(2) and any structure $\mathfrak{B}$

$$\text{if } \bar{a} \in A, \bar{b} \in B, (\mathfrak{A},\bar{a}) \simeq_{n_i} (\mathfrak{B},\bar{b}) \text{ and } \sigma = \widetilde{\varphi_i^{\mathfrak{A}}(\cdot,-,\bar{a})},$$

$$\text{then } (\mathfrak{B},\widetilde{\varphi_i^{\mathfrak{B}}(\cdot,-,\bar{b})}) \models \varphi.$$

Now, fix any linear ordering on $A = \{1,\ldots,r\} \times \{1,\ldots,n\}$. Choose as χ in (ii) a formula $\chi(\bar{w}_1,\ldots,\bar{w}_n)$ expressing

" given $i \in \{1,\ldots,r\}$ and any $\bar{w}$, if <u>bas</u>$^{\varphi_i(\cdot,-,\bar{w})}$ and $\varphi^{\varphi_i(\cdot,-,\bar{w})}$

then for some $(j,l) \in A$, $\widetilde{\varphi_i(\cdot,-,\bar{w})} = \widetilde{\varphi_j(\cdot,-,\bar{w}_l)}$ " .

Let $\Theta_i(x,\bar{y},\bar{w}_1,\ldots,\bar{w}_n)$ be a formula expressing

" for all $(k,l) \in A$, if (k,l) is the member in the ordering of A leading to the i-th distinct topology which is a model of φ, then $\varphi_k(x,\bar{y},\bar{w}_l)$".

Suppose that L resp. L' is a similarity type appropriate for structures $(\mathfrak{A},\sigma)$ resp. $(\mathfrak{A},\sigma,\tau)$. Suppose that T is an L'_t-theory that defines implicitly the second topology, i.e.

if $(\mathfrak{A},\sigma,\tau_1) \models T$ and $(\mathfrak{A},\sigma,\tau_2) \models T$ then $\tau_1 = \tau_2$.

Later on we will show that in general the second topology is not definable in T by an L_t-formula (compare also 8.8.6). But there are also some positive results in this case

(a) If T defines implicitly the second topology, then there is a $T^* \subset L_t$ such that for all $(\mathfrak{A},\sigma)$,

$(\mathfrak{A},\sigma) \models T^*$ iff $(\mathfrak{A},\sigma,\tau) \models T$ for some topology τ.

(b) Let $L(\dot{I})$ be the logic for topological structures $(\mathfrak{A},\sigma)$ obtained from $L_{\omega\omega}$ adding a new formation rule for the logical symbol I:

if φ is a formula, then $Itx\varphi$ is a formula

(x is a bounded variable of $Itx\varphi$) and adding the following clause in the definition of the satisfaction relation (assume $\varphi = \varphi(x,\bar{y})$ $t = t(\bar{y})$)

$(\mathfrak{A},\sigma) \models Itx\varphi[\bar{a}]$ iff $t^{\mathfrak{A}}[\bar{a}]$ is an interior point of $\{b \mid (\mathfrak{A},\sigma) \models \varphi[b,\bar{a}]\}$.

Let $L(I)_t$ be a language for structures $(\mathfrak{A},\sigma,\tau)$ (σ,τ being topologies on A) having the symbol I for the first topology and the usual second-order quantifiers of L_t for the second topology. If T defines implicitly the second topology, then some $\varphi(x,\bar{y}) \in L(I)$ explicitly defines this topology in T.

§ 8 Lindströms theorem and related logics.

First we prove a Lindström theorem for L_t, i.e. we show that there is no logic for topological structures stronger then L_t, which satisfies a compactness theorem and a Löwenheim-Skolem theorem. Then we introduce languages for uniform structures, for proximity spaces and for monotone structures and study their relations.

Since for topological structures the definitions of a logic, of the compactness property,... are the trivial extensions of the corresponding notions for "classical" model theory, we sketch them briefly.

A logic $\mathcal{L}$ for topological structures is a pair $(\mathcal{L},\models_{\mathcal{L}})$, where $\mathcal{L}$ is a function which associates to any many-sorted similarity type L a class $L_{\mathcal{L}}$ - the class of L-sentences of $\mathcal{L}$; and $\models_{\mathcal{L}}$ is a binary relation: if $\mathfrak{A} \models_{\mathcal{L}} \varphi$, then for some

L, $\mathfrak{A}$ is a topological L-structure and $\varphi \in L_{\mathcal{L}}$. We then say $\mathfrak{A}$ is a model of φ. We assume that $\underline{\mathcal{L}}$ satisfies some basic properties (e.g. homeomorphic structures are models of the same $\underline{\mathcal{L}}$-sentences, $\underline{\mathcal{L}}$ has a renaming property, $\underline{\mathcal{L}}$ is closed under Boolean operations,...).

Let $\underline{\mathcal{L}}_t$ be the logic for topological structures given by the language L_t. Further examples of logics are given by $L_2, (L_{\omega_1\omega})_t, L_t(Q)$.

We say that $\underline{\mathcal{L}}$ satisfies the compactness theorem, if any set of $\underline{\mathcal{L}}$-sentences has a model whenever each finite subset does. $\underline{\mathcal{L}}$ satisfies the Löwenheim-Skolem theorem, if any denumerable set of $\underline{\mathcal{L}}$-sentences, that is satisfiable, has a model in which all universes are denumerable and all topologies have a denumerable basis.

Given logics $\underline{\mathcal{L}}_1$ and $\underline{\mathcal{L}}_2$ for topological structures, we say that $\underline{\mathcal{L}}_1$ is as strong as $\underline{\mathcal{L}}_2$, and write $\underline{\mathcal{L}}_1 \geq \underline{\mathcal{L}}_2$, if for every $\underline{\mathcal{L}}_2$-sentence there is an $\underline{\mathcal{L}}_1$-sentence with the same models. We say that $\underline{\mathcal{L}}_1$ is stronger than $\underline{\mathcal{L}}_2$, if $\underline{\mathcal{L}}_1 \geq \underline{\mathcal{L}}_2$ but not $\underline{\mathcal{L}}_2 \geq \underline{\mathcal{L}}_1$.

8.1 Theorem. Let $\underline{\mathcal{L}}$ be a logic for topological structures with $\underline{\mathcal{L}} \geq \underline{\mathcal{L}}_t$. If $\underline{\mathcal{L}}$ satisfies the compactness theorem and the Löwenheim-Skolem theorem, then $\underline{\mathcal{L}}$ is not stronger than $\underline{\mathcal{L}}_t$.

Proof. The proof is essentially the same as the proof of Lindströms theorem for $L_{\omega\omega}$, so that we only sketch the main points of it.

For $\varphi \in L_{\mathcal{L}}$ we have to show that there is a $\chi \in L_t$ with the same models. For simplicity assume that L is one-sorted. By an application of the compactness theorem for $\underline{\mathcal{L}}$ one can assume that φ is (essentially) an $L_{\mathcal{L}}$-sentence for some finite L.

Recall that in 4.5 and 4.11 we defined sets of sentences

$$\Phi_o \subset \Phi_1 \subset \ldots \qquad \text{and} \qquad \Phi_p = \bigcup_{n \geq o} \Phi_n$$

such that

$$(\mathfrak{B}_1, \tau_1) \simeq_n^t (\mathfrak{B}_2, \tau_2) \quad \text{iff} \quad ((\mathfrak{B}_1, \tau_1), (\mathfrak{B}_2, \tau_2), \ldots) \models \Phi_n \text{ for some} \ldots,$$

$$(\mathfrak{B}_1, \tau_1) \simeq_p^t (\mathfrak{B}_2, \tau_2) \quad \text{iff} \quad ((\mathfrak{B}_1, \tau_1), (\mathfrak{B}_2, \tau_2), \ldots) \models \Phi_p \text{ for some} \ldots.$$

For $n \geq 0$ put

$$\psi^n = \bigvee\{\varphi^n_{(\mathfrak{A},\sigma)} \mid (\mathfrak{A},\sigma) \models_{\mathcal{L}} \varphi\}$$

(for the definition of $\varphi^n_{(\mathfrak{A},\sigma)}$ see section 4).

Since $\varphi \models_{\mathcal{L}} \psi^n$ and $\models_t \psi^{n+1} \to \psi^n$, it suffices to show that

$$\{\psi^n \mid n \in \omega\} \models_{\mathcal{L}} \varphi.$$

Thus, assume $(\mathfrak{B},\tau)$ is a model of $\{\psi^n \mid n \in \omega\}$. Then, for each $n \in \omega$, there is a substructure $(\mathfrak{A}_n,\sigma_n)$ such that

$$(\mathfrak{A}_n,\sigma_n) \models_{\mathcal{L}} \varphi \quad \text{and} \quad (\mathfrak{A}_n,\sigma_n) \simeq^t_n (\mathfrak{B},\tau) .$$

Hence for each n and some ...

$$((\mathfrak{A}_n,\sigma_n),(\mathfrak{B},\tau),\ldots) \models \Phi_n .$$

If $(\mathfrak{B},\tau)$ is not a model of φ, then we find, using the compactness theorem and the Löwenheim-Skolem theorem for $\underline{\mathcal{L}}$, structures $(\mathfrak{A}^*,\sigma^*)$ and $(\mathfrak{B}^*,\sigma^*)$ such that

(1) $((\mathfrak{A}^*,\sigma^*),(\mathfrak{B}^*,\tau^*),\ldots) \models \Phi_p$.

(2) $(\mathfrak{A}^*,\sigma^*) \models_{\mathcal{L}} \varphi$, $(\mathfrak{B}^*,\tau^*)$ is a model of the negation of φ.

(3) A^* and B^* are denumerable and σ^* and τ^* have denumerable bases.

By (1), $(\mathfrak{A}^*,\sigma^*)$ and $(\mathfrak{B}^*,\tau^*)$ are partially homeomorphic and hence by (3), homeomorphic, but this contradicts (2).

8.2 <u>Remarks and exercises.</u>

1) Note that the proof essentially contains the arguments of the proof of the convergence 4.16 (there we didn't need the sets Φ_n und Φ_p, since each denumerable set of <u>L_t-sentences</u> has a denumerable recursively saturated model and so we could apply 4.7). Some consequences of the convergence lemma may be viewed as applications of 8.1, e.g.: let $\underline{\mathcal{L}}$ be the logic for topological structures having as L-sentences the set of L_2-sentences invariant for topologies. Since $\underline{\mathcal{L}} \geq \mathcal{L}_t$ and $\underline{\mathcal{L}}$ satisfies the compactness theorem and the Löwenheim-Skolem theorem, we obtain from 8.1: each L_2-sentence invariant for topologies is equivalent to an L_t-sentence (see 4.19). Similarly one can derive the L_t-interpolation theorem using a more careful formulation of

8.1.

2) Let $\mathcal{L}$ be a logic with $\mathcal{L} \geq \mathcal{L}_t$. Suppose that each denumerable set of $\mathcal{L}$-sentences has a model in which all universes are denumerable (but nothing is required of the topologies!). Show that $\mathcal{L}$ satisfies the Löwenheim-Skolem theorem (compare the hint in 4.5).

3) Call a weak structure $(\mathfrak{A},\sigma)$ closed if $\sigma = \tilde{\sigma}$, i.e. if σ is closed under unions. Similarly for many-sorted weak structures. Let Φ be any fixed set of L_t-sentences for $L = \emptyset$. We restrict to many-sorted closed structures, where all sorts are models of Φ, and look at logics for this class of structures. Once more denote by $\mathcal{L}_t$ the logic induced by the language L_t. Prove the analogue of 8.1, i.e. show that there is no logic for closed models of Φ stronger than $\mathcal{L}_t$ and still satisfying the compactness theorem and the Löwenheim-Skolem theorem.- For $\Phi = \{\underline{bas}\}$ we obtain 8.1, for $\Phi = \{\underline{disc}\}$ this is essentially Lindströms theorem for $L_{\omega\omega}$. For $\Phi = \emptyset$ we obtain that $\mathcal{L}_t$ is a "maximal" logic for closed structures and for $\Phi = \{\underline{haus}\}$ we get that $\mathcal{L}_t$ is also maximal if we restrict to topological structures carrying Hausdorff topologies.

4) Show that there are logics $\mathcal{L}$ for topological structures satisfying the compactness theorem and the Löwenheim-Skolem theorem and such that neither $\mathcal{L} \geq \mathcal{L}_t$ nor $\mathcal{L}_t \geq \mathcal{L}$. Hint: Let $\mathcal{L}$ be the logic having for any L, L_t as set of L-sentences of $\mathcal{L}$, and where

$$(\mathfrak{A},\sigma) \models_{\mathcal{L}} \varphi \quad \text{iff} \quad \begin{cases} (\mathfrak{A},\sigma) \models_t \varphi & \text{, if A is infinite} \\ (\mathfrak{A},\{A-U \mid U \in \sigma\}) \models_t \varphi & \text{, if A is finite} \end{cases}$$

(and similarly for many-sorted structures).

Show that the $\mathcal{L}$-sentence $\forall x(Px \rightarrow \exists X \ni x\ \forall y(y \in X \rightarrow Py))$ is not equivalent to an L_t-sentence. Show that also for $\mathcal{L}$ a Lindström theorem holds.

We introduce now a language appropriate for the so called monotone structures. This language for example is useful for the study of uniform spaces and for the study of such topological structures, where the topology is determined by the set of neighborhoods of some fixed point (e.g. topological groups, fields,.....).

Assume throughout that k is a fixed natural number ≥ 1. Given a set A, we call a non-empty set μ of subsets of A^k, $\mu \subset P(A^k)$, a monotone system, if

$$B \in \mu \quad \text{and} \quad B \subset C \subset A^k \quad \text{imply} \quad C \in \mu .$$

Given a non-empty set $\beta \subset P(A^k)$, let

$$\hat{\beta} = \{C \mid B \subset C \subset A^k \quad \text{for some } B \in \beta\}.$$

Clearly, $\hat{\beta}$ is the least monotone system containing β.

$(\mathfrak{A},\mu)$ is called a monotone L-structure, if $\mathfrak{A}$ is an L-structure and μ is a monotone system. $(\mathfrak{A},\mu),(\mathfrak{B},\upsilon)$ will always be monotone structures.

8.3 Examples. (i) (k = 2) If μ is a uniformity on A (i.e. if (A,μ) is a uniform space), then (A,μ) is a monotone structure. (ii) (k = 1) If $(\mathfrak{A},\sigma)$ is a topological group and μ_σ is the set of neighborhoods of the unit in $\mathfrak{A}$, then $(\mathfrak{A},\mu_\sigma)$ is a monotone structure.

Let L_2^k be the second-order logic defined as L_2, the set variables $W_0, W_1, \ldots$ now being variables for k-ary relations. Thus L_2^k has besides the atomic formulas of $L_{\omega\omega}$ the atomic formulas $(t_1,\ldots,t_k) \in X$. Sometimes we write $Xt_1\ldots t_k$ for $(t_1,\ldots,t_k) \in X$.

8.4 Definition. An L_2^k-sentence is called invariant (more precisely, invariant for monotone structures), if for any $(\mathfrak{A},\beta)$ we have

$$(\mathfrak{A},\beta) \models \varphi \quad \text{iff} \quad (\mathfrak{A},\hat{\beta}) \models \varphi .$$

Thus, when restricting to monotone structures and invariant L_2^k-sentences, the compactness theorem and the Löwenheim-Skolem theorem hold.

We denote by L_m^k the set of L_2^k-formulas obtained from the atomic formulas by the formation rules of $L_{\omega\omega}$ and the rules:

(i) If φ is positive in X, then $\forall X\varphi$ is a formula.

(ii) If φ is negative in X, then $\exists X\varphi$ is a formula.

For example, the class of structures $(\mathfrak{A},\mu)$ where μ is a uniformity on A is the class of monotone models of the following set Φ_u of L_2^2-sentences:

$$\Phi_u = \{\forall X\, \forall x\, Xxx, \forall X\, \forall Y\, \exists Z\, \forall x\, \forall y(Zxy \rightarrow (Xxy \wedge Yxy)),$$
$$\forall X\, \exists Y\, \forall x\, \forall y\, \forall z(Yyx \wedge Yyz \rightarrow Xxz)\}.$$

One shows by induction (compare 2.3)

8.5 Lemma. Every L^k_m-sentence is invariant for monotone structures. The converse of 8.5 is obtained in the same way as the corresponding result for L_t in section 4:

Note that $\varphi \in L^k_2$ is invariant iff it is preserved under the relation M, where

$$(\mathfrak{A}, \beta)M(\mathfrak{B}, \gamma) \quad \text{iff} \quad (\mathfrak{A}, \hat{\beta}) \text{ and } (\mathfrak{B}, \hat{\gamma}) \text{ are isomorphic.}$$

In particular, we have (for k = 1)

$(\mathfrak{A}, \beta)M(\mathfrak{B}, \gamma)$ iff there are a map π^o and relations π^1, π^2, $\pi^1, \pi^2 \subset \beta \times \gamma$ such that (1), (2) as formulated at the beginning of § 4 hold and
(3) for every $V \in \gamma$ there is some $U \in \beta$ with $U \pi^1 V$
(4) for every $U \in \beta$ there is some $V \in \gamma$ with $U \pi^2 V$.

Now it should be clear how one can define the notion of partial isomorphism, the back and forth properties.. and how one can derive for L^k_m the results corresponding to those of § 4 - § 7. In particular, since the formulas defining the finite approximations of the relation M are L^k_m-formulas, one obtains:

8.6 Theorem. Each L^k_2-sentence invariant for monotone structures is equivalent to an L^k_m-sentence. - More generally: Let $\Phi \subset L^k_m$. If φ is invariant for models of Φ (i.e. if $(\mathfrak{A}, \beta) \models \Phi$ implies $((\mathfrak{A}, \beta) \models \varphi$ iff $(\mathfrak{A}, \hat{\beta}) \models \varphi))$, then $\Phi \models \varphi \leftrightarrow \psi$ for some $\psi \in L^k_m$.

8.7 Theorem. Let Φ be a set of L^k_m-sentences. We restrict to monotone structures that are models of Φ. Then there is no logic for this class of structures stronger than L^k_m and still satisfying the compactness theorem and the Löwenheim-Skolem theorem.

8.8 Remarks, examples and exercises.

1) (Uniform spaces) a) Taking Φ_u (the set of L^2_m-sentences axiomatizing uniform structures, see above) as Φ in 8.6 and 8.7, we see that the invariant sentences for uniform structures are - up to the equivalence - the L^2_m-sentences and that L^2_m is a "maximal" logic for uniform structures.

b) Given a uniform structure $(\mathfrak{B}, \mu)$ denote by τ_μ the topology induced by μ.

Show that for any $\varphi \in L_t$ there is a $\psi \in L_m^2$ such that for any uniform structure $(\mathfrak{A},\mu)$,

$$(\mathfrak{A},\mu) \models \psi \quad \text{iff} \quad (\mathfrak{A},\tau_\mu) \models \varphi .$$

Show that there are uniform structures $(\mathfrak{A},\mu_1)$ and $(\mathfrak{A},\mu_2)$ with discrete $\tau_{\mu_1} = \tau_{\mu_2}$ but not satisfying the same L_m^2-sentences.

2) Given any topological group $(\mathfrak{A},\sigma)$ denote by μ_σ the monotone system on A (for k = 1) consisting of the neighborhoods of the unit in $\mathfrak{A}$. Show that for any $\varphi \in L_t$ there is a $\varphi' \in L_m$ and that for any $\psi \in L_m$ there is a $\psi' \in L_t$ such that for any topological group

$$(\mathfrak{A},\sigma) \models \varphi \quad \text{iff} \quad (\mathfrak{A},\mu_\sigma) \models \varphi'$$

and

$$(\mathfrak{A},\sigma) \models \psi' \quad \text{iff} \quad (\mathfrak{A},\mu_\sigma) \models \psi .$$

3) ("The class of topological structures forms a definable class of montone structures".) Given a topology σ on A let σ^* be the monotone system on A x A (i.e. k = 2) generated by

$$\{\{a\} \times U | a \in A, U \in \sigma \text{ and } a \in U\}.$$

Since V is a neighborhood of a iff $\{a\} \times V \in \sigma^*$, we have:

a) If σ and τ are topologies on A with $\sigma^* = \tau^*$, then $\sigma = \tau$.

b) There is a sentence <u>top</u> $\in L_m^2$ (for $L = \emptyset$) such that for any monotone structure $(\mathfrak{A},\mu)$,

$$(\mathfrak{A},\mu) \models \underline{top} \quad \text{iff} \quad \mu = \sigma^* \text{ for some topology } \sigma \text{ on A}.$$

c) For any $\varphi \in L_m^2$ there is a $\check{\varphi} \in L_t$ such that for any topological structure $(\mathfrak{A},\sigma)$,

$$(\mathfrak{A},\sigma) \models \check{\varphi} \quad \text{iff} \quad (\mathfrak{A},\sigma^*) \models \varphi .$$

d) For any $\varphi \in L_t$ there is a $\hat{\varphi} \in L_m^2$ such that for any topological structure $(\mathfrak{A},\sigma)$,

$$(\mathfrak{A},\sigma) \models \varphi \quad \text{iff} \quad (\mathfrak{A},\sigma^*) \models \hat{\varphi} .$$

To show c resp. d replace set quantifiers in φ as indicated by

$$\exists X \ldots Xt_1t_2 \dashrightarrow \exists x\, \exists X \ni x \ldots t_1 = x \wedge t_2 \in X \text{ ---}$$

resp.

$$\exists X \ni t \ldots t_1 \in X \dashrightarrow \exists X (\forall x\, \forall z (Xxz \rightarrow x = t) \wedge \ldots Xtt_1 \text{ --- }).$$

4) Since L_m^k contains only "unbounded" quantifiers on set variables, it is sometimes easier to deal with L_m^k than with L_t. In many cases one can then translate the result to L_t using c and d of the preceding remark. As an example we sketch a syntactic proof of the interpolation theorem for L_m^k: For simplicity take $k = 1$ and write L_m for L_m^1. For a countable set C of new constants and a countable set $\mathfrak{U}$ of new set constants let $L(C,\mathfrak{U})_2$ be defined as in 1.4 and denote by $L(C,\mathfrak{U})_m$ the set of "monotone" formulas of $L(C,\mathfrak{U})_2$. A <u>sequent</u> S is a finite set of $L(C,\mathfrak{U})_2$-sentences in negation normal form. We say that a sequent S_1 is <u>valid</u> ,and write $\models S_1$ if $\models \bigvee S_1$, i.e. if any weak structure is a model of at least one sentence in S_1. We say that S_1 is <u>derivable</u> and write $\vdash S_1$, if S_1 is derivable using the following axioms and rules:

Axioms: $S,\varphi,\neg\varphi$ where φ is atomic (and $S,\varphi,\neg\varphi$ denotes the sequent $S \cup \{\varphi\} \cup \{\neg\varphi\}$).

Rules: $(\wedge)$ $\dfrac{S,\varphi \quad S,\psi}{S,(\varphi\wedge\psi)}$ $\qquad (\vee)$ $\dfrac{S,\varphi}{S,(\varphi\vee\psi)}$ $\qquad \dfrac{S,\psi}{S,(\varphi\vee\psi)}$

$(\forall x)$ $\dfrac{S,\varphi\frac{c}{x}}{S,\forall x\varphi}$ $\qquad (\forall X)$ $\dfrac{S,\varphi\frac{U}{X}}{S,\forall X\varphi}$

where c resp. U does not occur in the conclusion

$(\exists x)$ $\dfrac{S,\varphi\frac{c}{x}}{S,\exists x\varphi}$ $\qquad (\exists X)$ $\dfrac{S,\varphi\frac{U}{X}}{S,\exists X\varphi}$

$(=_1)$ $\dfrac{S,\neg c = c}{S}$ $\qquad (=_2)$ $\dfrac{S,\neg fc_1\ldots c_n = c}{S}$ where c does not occur in $S, fc_1\ldots c_n$.

$(=_3)$ $\dfrac{S,\varphi\frac{c}{x}}{S,\varphi\frac{t}{x},\neg t = c}$ where φ is atomic or the negation of an atomic formula,and t is a basic term.

a) Prove the completness theorem, i.e. show that for any sequent S, $\models S$ iff $\vdash S$.

b) Suppose that S^1 and S^2 are sequents and that $S^i \subset L^i(C,\mathfrak{U})_m$ for $i = 1,2$. Let $L^o = L^1 \cap L^2$. Give a syntactic proof (i.e. a proof on induction on the length of the derivation of $\vdash S^1,S^2$) of the following:

If $\vdash S^1, S^2$, then there is a $\chi \in L^o(C,\mathfrak{U})_m$ such that

(i) $\vdash S^1, \chi$ and $\vdash S^2, \neg\chi$

(ii) if $U \in \mathfrak{U}$ occurs negatively in χ, then U occurs positively in S^1,
if $U \in \mathfrak{U}$ occurs positively in χ, then U occurs positively in S^2.

From this derive the interpolation theorem for L_m.

c) Generalize a and b to many-sorted languages, thus obtaining, as indicated in 5.2, a syntactic proof of the fact that the L_2-sentences invariant for monotone structures are the L_m-sentences.

d) From b and c derive, using c and d of the preceding remark 8.8.3, the interpolation theorem for L_t and the characterization of L_t as set of L_2-sentences invariant for topologies. - Considering an axiomatization with appropriate rules for bounded quantifiers (similar to that in [4]), it is possible to give a direct syntactic proof of these results for L_t.

5) The "natural" generalization to L_m $(= L^1_m)$ of the Chang-Makkai theorem for relations does not hold, i.e. there is an $(L \cup \{R\})_m$-theory T such that

(i) for any denumerable L-structure $(\mathfrak{A},\mu)$ (i.e. A is denumerable and μ has a denumerable basis) the set $\{R^A \mid ((\mathfrak{A},R^A),\mu) \models T\}$ is at most countable.

(ii) there is no finite set of L_m-formulas $\varphi_1(x,\bar{y}),\dots,\varphi_r(x,\bar{y})$ such that

$$T \models \bigvee_{i=1}^{r} \exists\bar{y}\ \forall x(Rx \leftrightarrow \varphi_i(x,\bar{y})).$$

Take as T the $\{R\}_m$-theory saying that R is a minimal set of the monotone system, i.e.

$$T = \{\exists X\ \forall x(Xx \rightarrow Rx), \forall X\ \forall y(Ry \rightarrow Xy)\}\ .$$

Clearly T satisfies (i). To show that (ii) holds take as μ a monotone system on $\mathbb{N}$ generated by a set of $\aleph_1$-many pairwise incomparable subsets of $\mathbb{N}$.

6) Look at structures $(\mathfrak{A},\mu_1,\mu_2)$ where μ_1 and μ_2 are monotone systems on A (for symplicity assume k = 1, i.e. $\mu_1,\mu_2 \subset P(A)$). Let L_m be the corresponding "monotone" language. Let $X, X_1, \dots$ denote set variables of the first sort, $Y, Y_1, \dots$ set variables of the second sort. Suppose $\varphi(x)$ is an L_m-formula containing only set variables of the first sort. Assume that φ has the form $\Pi \neg \psi$, where Π is a prefix (possibly containing individual and set

variables) and ψ is an L_m-formula.

Suppose T is an L_m-theory such that

$$T \models \forall Y\, \Pi\, \forall x(\psi \rightarrow x \in Y) \quad \text{and} \quad T \models \neg\, \Pi\, \forall Y\, \exists x(x \in Y \wedge \neg\, \psi).$$

Show: a) T defines implicit the second monotone system, namely if $(\mathfrak{A},\mu_1,\mu_2) \models T$ then $\mu_2 = \{B \mid (\mathfrak{A},\mu_1) \models \Pi\, \forall x(\psi \rightarrow x \in B)\}$.

b) There is a set T* of L_m-sentences containing only set variables of the first sort such that for all $(\mathfrak{A},\mu)$

$$(\mathfrak{A},\mu) \models T^* \quad \text{iff} \quad (\mathfrak{A},\mu,\upsilon) \models T \quad \text{for some } \upsilon .$$

Show that the "natural" generalization to L_t of the Chang-Makkai theorem for relations does not hold. (Hint: Look at structures $(\mathfrak{A},\mu_1,\mu_2)$ where $\mu_1 \subset A^2$ and $\mu_2 \subset A$ are monotone systems, μ_1 is closed under intersections and $\mu_2 = \{B \mid (\mathfrak{A},\mu_1) \models \forall X\, \exists y\, \forall x(\neg\, Uxy \rightarrow x \in B)\}$), and we use b and 5).

7) Generalizing topological and monotone structures, look at structures that have attached to each point of its universe a monotone system, i.e. structures $(\mathfrak{A},\lambda)$ where λ is a function defined on A, and for each $a \in A$, $\lambda(a)$ is a monotone system on A (i.e. $\lambda(a) \subset P(A)$ and $\lambda(a)$ is monotone). We call such structures <u>point-monotone structures.</u>

If $(\mathfrak{A},\sigma)$ is a topological structure, let $(\mathfrak{A},\lambda_\sigma)$ be the point-monotone structure defined by

$$\lambda_\sigma(a) = \{B \mid B \text{ is a neighborhood of } a\}.$$

If $\mathfrak{A}$ is an L-structure and $< \in L$, let $(\mathfrak{A},\lambda_{<^A})$ be the point-monotone structure given by

$$\lambda_{<^A}(a) = \{B \mid \{b \mid b <^A a\} \subset B \subset A\}.$$

The language L_{2*} for point-monotone structures is built with the same symbols as L_2. L_{2*}-formulas beginning with set quantifiers are obtained by the rule:

if φ is a formula, X a set variable and t a term, then
$\forall X(t)\varphi$ and $\exists X(t)\varphi$ are formulas.

Define

$$(\mathfrak{A},\lambda) \models \forall X(t)\ \varphi(X) \quad \text{iff} \quad (\mathfrak{A},\lambda) \models \varphi[B] \quad \text{for all } B \in \lambda(t^{\mathfrak{A}}),$$

and do similarly for $\exists X(t)\ \varphi(X)$.

Define the notion of an L_{2*}-sentence invariant for point-monotone structures. Let L_{pm} be the set of L_{2*}-formulas containing a quantifier $\forall X(t)\ \varphi(X)$ (resp. $\exists X(t)\ \varphi(X)$) only if φ is positive in X (resp. φ is negative in X).

a) Show with the corresponding back and forth technique that the invariant L_{2*}-sentences are, up to equivalence, the L_{pm}-sentences.

Call $(\mathfrak{B},\lambda')$ an extension of $(\mathfrak{A},\lambda)$, if $\mathfrak{A} \subset \mathfrak{B}$ and for any $a \in A$, $\lambda(a) = \{A \cap C \mid C \in \lambda'(a)\}$. If furthermore $\lambda(a) \subset \lambda'(a)$ for all $a \in A$, then $(\mathfrak{B},\lambda')$ is called an <u>open extension</u> of $(\mathfrak{A},\lambda)$. Show

b) If $(\mathfrak{A},\sigma)$ and $(\mathfrak{B},\tau)$ are topological structures, then

$(\mathfrak{B},\lambda_\tau)$ is an open extension of $(\mathfrak{A},\lambda_\sigma)$ iff $(\mathfrak{B},\tau)$ is an open extension of $(\mathfrak{A},\sigma)$ in the sense of § 5.

If $\mathfrak{A}$ and $\mathfrak{B}$ are L-structures and $< \in L$, then

$(\mathfrak{B},\lambda_{<^B})$ is an open extension of $(\mathfrak{A},\lambda_{<^A})$ iff $\mathfrak{B}$ is an end-extension of $\mathfrak{A}$.

c) Let T be a set of L_{pm}-sentences. If $\varphi \in L_{pm}$ is closed under open extensions, when restricting to models of T, then there is a $\psi \in L_{pm}$ in negation normal form such that $T \models \varphi \leftrightarrow \psi$ and any subformula of ψ beginning with a universally quantified variable has the form $\forall x(x \in Y \rightarrow \psi)$.

d) Let T_o be an $L_{\omega\omega}$-theory and assume $< \in L$. Suppose that $\varphi \in L_{\omega\omega}$ is closed under end-extensions for models of T_o. Choose $T \subset L_{pm}$ such that

$$(\mathfrak{A},\lambda) \models T \quad \text{iff} \quad \mathfrak{A} \models T_o \quad \text{and} \quad \lambda = \lambda_{<^A}.$$

Then φ is closed under open extensions for models of T. Choose $\psi \in L_{pm}$ as in c, i.e. $T \models \varphi \leftrightarrow \psi$, ψ is in negation normal form and has only bounded universal quantifiers on individual variables. Let $\check{\psi}$ be the $L_{\omega\omega}$-sentence obtained from ψ replacing set quantifiers as indicated by

$$\forall X(t) \ldots t' \in X \text{---} \quad \rightarrow \quad \ldots t' < t \text{---}$$

$$\exists X(t) \ldots t' \in X \text{---} \quad \rightarrow \quad \ldots t' < t \text{---} .$$

Then $\check{\psi}$ is restricted existential and $T_o \models \varphi \leftrightarrow \check{\psi}$.

e) Similarly derive from c the result 5.12 for sentences $\varphi \in L_t$ preserved by open extensions.

8) (Proximity-spaces) [16] (A,δ) is called a proximity space if A is a non-empty set and δ is a binary relation on P(A), i.e. $\delta \subset P(A) \times P(A)$, satisfying (i) - (vi): we write $B\delta C$ for $(B,C) \in \delta$ and $B\not\delta C$ for $(B,C) \notin \delta$; read $B\delta C$ (resp. $B\not\delta C$) as "B and C are distant" (resp. "B and C are proximate")

(i) if $B\delta C$ then $C\delta B$

(ii) if $B\delta C$ and $B' \subset B$ then $B'\delta C$

(iii) if $B_1\delta C$ and $B_2\delta C$ then $B_1 \cup B_2\delta C$

(iv) for every $a \in A$, $\{a\}\not\delta\{a\}$

(v) $\emptyset\delta A$

(vi) if $B\delta C$ then there are B',C' such that $B \subseteq B'$, $C \subseteq C'$, $B' \cap C' = \emptyset$ and $B\delta(A - B')$ and $(A-C')\delta C$.

Call $(\mathfrak{A},\delta)$ a proximity structure if (A,δ) is a proximity space.

Given an arbitrary non-empty $\delta \subset P(A) \times P(A)$, denote by $\bar{\delta}$ the set

$$\bar{\delta} = \{(B,C) \mid B \subset B' \text{ and } C \subset C' \text{ for some } (B',C') \in \delta\}.$$

If (A,δ) is a proximity space then, by (i) and (ii), $\delta = \bar{\delta}$.

We introduce a second-order language L_{2o} appropriate for structures $(\mathfrak{A},\delta)$. This language contains, besides the individual variables, the "pairs" of second-order variables $(W_1^{\ 1},W_2^{\ 1}),(W_1^{\ 2},W_2^{\ 2})\ldots$ .We denote them by (X_1,X_2), $(Y_1,Y_2),\ldots$ .Besides the $L_{\omega\omega}$-atomic formulas we have in L_{2o} for any term t and any variable (X_1,X_2) the atomic formulas $t \in X_1$ and $t \in X_2$. L_{2o} is closed under the formation rules of $L_{\omega\omega}$ and:

if φ is a formula then $\forall(X_1,X_2)\ \varphi$ and $\exists(X_1,X_2)\ \varphi$ are formulas.

Let prox be the conjunction of the following L_{2o}-sentences:

$$\varphi_{(i)} = \forall(X_1,X_2)\ \exists(Y_1,Y_2)\ (\text{"}X_2 \subset Y_1\text{"} \wedge \text{"}X_1 \subset Y_2\text{"})\ .$$

$$\varphi_{(iii)} = \forall(X_1,X_2)\ \forall(Y_1,Y_2)\ \exists(Z_1,Z_2)\ (\text{"}X_1 \cup Y_1 \subset Z_1\text{"} \wedge \text{"}X_2 \cap Y_2 \subset Z_2\text{"})$$

$$\varphi_{(iv)} = \forall(X_1,X_2)\ \forall x(\neg\, x \in X_1 \vee \neg\, x \in X_2)$$

$\varphi_{(v)} = \exists(X_1,X_2)\ \forall x\ x \in X_2$

$\varphi_{(vi)} = \forall(X_1,X_2)\ \exists(Y_1,Y_2)\ \exists(Z_1,Z_2)(\forall x(x \in Y_2 \vee x \in Z_2) \wedge "X_1 \subset Y_1"$

$\wedge "X_2 \subset Z_2").$

a) Show:

For any $(\mathfrak{A},\delta)$ we have

$(\mathfrak{A},\delta) \models \underline{\text{prox}}$ iff $(\mathfrak{A},\bar{\delta})$ is a proximity structure.

Let L_p be the set of L_{2o}-formulas containing a quantifier $\forall(X_1,X_2)\varphi$ (resp. $\exists(X_1,X_2)\varphi$) only if φ is negative in X_1 and X_2 (resp. φ is postive in X_1 and X_2). Show:

b) prox "is" an L_p-sentence.

c) L_p-sentences are invariant, i.e. if $\varphi \in L_p$ then for any $(\mathfrak{A},\delta)$,

$(\mathfrak{A},\delta) \models \varphi$ iff $(\mathfrak{A},\bar{\delta}) \models \varphi.$

d) When restricting to proximity structures, L_p is a logic satisfying the compactness and the Löwenheim-Skolem theorem.

In the following we restrict us to structures $(\mathfrak{A},\delta)$ with $(\emptyset,A) \in \delta$ and $(A,\emptyset) \in \delta$. - Given (A,δ) let $\beta_\delta \subset P(A^2)$ be the system of sets $\beta_\delta = \{A^2 - B\times C \mid B\delta C\}$. Show

e) For any $(A,\delta_1),(A,\delta_2)$,

$\bar{\delta}_1 = \bar{\delta}_2$ iff $\hat{\beta}_{\delta_1} = \hat{\beta}_{\delta_2}$ (where $\hat{\beta}_{\delta_i}$ is the monotone system on A^2 generated by β_{δ_i}).

f) Let φ_o be the L^2_m-sentence

$\forall X\ \exists Y\ \forall x\ \forall y((\exists u \neg Xxu \wedge \exists v \neg Xvy) \rightarrow \neg Yxy).$

Then, for any monotone structure $(\mathfrak{A},\mu)$,

$(\mathfrak{A},\mu) \models \varphi_o$ iff $\mu = \hat{\beta}_\delta$ for some $\delta \subset P(A) \times P(A)$.

g) For any $\varphi \in L_p$ (resp. $\psi \in L_m$) there is a $\varphi^m \in L_m$ (resp. $\psi^p \in L_p$) such that for any $(\mathfrak{A},\delta)$,

$(\mathfrak{A},\delta) \models \varphi$ iff $(\mathfrak{A},\beta_\delta) \models \varphi^m$ and

$(\mathfrak{A},\delta) \models \varphi^p$ iff $(\mathfrak{A},\beta_\delta) \models \psi.$

In particular, the class of monotone structures which are a model of $\varphi_o \wedge \underline{\text{prox}}^m$ is

$$\{(\mathfrak{A},\hat{\beta}_\delta) \mid (\mathfrak{A},\delta) \text{ a proximity structure}\}.$$

h) Using f and g and the corresponding results for L_m show that the interpolation theorem holds for L_p, and when restricting to proximity structures, that L_p is a maximal logic satisfying the compactness theorem and the Löwenheim-Skolem theorem. In particular, the invariant L_{2^o}-sentences are, up to equivalence, the L_p-sentences.

i) Given a proximity space (A,δ) denote by σ_δ the topology induced by δ. Show that for any $\varphi \in L_t$ there is a $\psi \in L_p$ such that for all proximity structures $(\mathfrak{A},\delta)$,

$$(\mathfrak{A},\delta) \models \psi \quad \text{iff} \quad (\mathfrak{A},\sigma_\delta) \models \varphi .$$

§ 9 Omitting types theorem

If T is a complete $L_{\omega\omega}$-theory with denumerable L and $\Phi(x) = \{\varphi_i(x) \mid i \in \omega\}$ is a type of T, then the omitting types theorem for $L_{\omega\omega}$ tells us that Φ is realized in each model of T iff there is some $\psi(x)$ such that

(*) $$T \models \exists x\, \psi(x) \quad \text{and} \quad T \models \forall x(\psi(x) \rightarrow \varphi_i(x)) \text{ for all } i \in \omega.$$

We show that in case T and Φ consist of L_t-formulas, we cannot find, in general, a ψ in L_t satisfying (*), i.e. the omitting types theorem does not generalize to L_t. Then we prove an omitting types theorem for the fragment L(I) (see §. 7) of L_t.

One method to get a model of T omitting a type Φ consists in enlarging step by step T to a Hintikka set Ω in such a way that Ω contains for each constant $c \in C$ (C the set of Henkin constants) a formula $\neg \varphi_i \frac{c}{x}$. If after a finite number of steps it is not possible to carry on this process, then for some c and a finite number of L_t-formulas $\psi_1(c,c_o,\ldots,c_r,U_o,\ldots,U_s),\ldots,$ $\psi_n(c,c_o,\ldots,c_r,U_o,\ldots,U_s)$, we have

$$T \models_t \psi_1 \wedge \ldots \wedge \psi_n \rightarrow \varphi_i \frac{c}{x} \quad \text{for all } i \in \omega \quad ,$$

and $T \models_t \psi$

where $\psi = \exists x\, \exists X_o \ldots \exists X_s\, \exists x_o \ldots \exists x_r \bigwedge_{i=1}^{n} \psi_i(x, x_o, \ldots, x_r, X_o, \ldots, X_s)$.

But in general $\psi \notin L_t$, since a set variable X_j may occur positively in some ψ_i.

Note that in ψ_1 = "X is an partial injective function" the set variable X occurs negatively, and in ψ_2 = "the domain of X is to whole universe" X occurs positively. Therefore, $\exists X(\psi_1 \wedge \psi_2)$ is not an L_t-formula. We make use of this fact to obtain the counterexample presented here.

9.1 Theorem. There is a complete and countable L_t-theory T and a set $\Phi(x) = \{\varphi_i(x) \mid i \in \omega\}$ of L_t-formulas such that

a) Each model of T realizes $\Phi(x)$.

b) There is no $\psi(x) \in L_t$ such that

$$T \models_t \exists x\, \psi(x) \quad \text{and} \quad T \models_t \forall x\, (\psi(x) \to \varphi_i(x)) \quad \text{for all } i \in \omega.$$

Proof. Take $L = \{M, A, B, S, Q, R, f, c\}$ where M, A, B, S are unary and Q, R, f are binary.

Let $(\mathfrak{C}, \tau)$ be the topological L-structure given by

$C = M^{\mathfrak{C}} \,\dot\cup\, A^{\mathfrak{C}} \,\dot\cup\, B^{\mathfrak{C}} \,\dot\cup\, S^{\mathfrak{C}} \,\dot\cup\, \{c^{\mathfrak{C}}\}$ with

$$M^{\mathfrak{C}} = \mathbb{N}, \quad A^{\mathfrak{C}} = \dot\bigcup_{n \in \mathbb{N}} A_n, \quad B^{\mathfrak{C}} = \dot\bigcup_{n \in \mathbb{N}} B_n, \quad S^{\mathfrak{C}} = \dot\bigcup_{n \in \mathbb{N}} (A_n \times B_n),$$

and where, for all $n \in \mathbb{N}$, $|A_n| = \aleph_o$, $|B_{2n}| = \aleph_o$, and $|B_{2n+1}| = n+1$.

$Q^{\mathfrak{C}} = \{(n,a) \mid a \in A_n\}$, $R^{\mathfrak{C}} = \{(n,b) \mid b \in B_n\}$,

$$f^{\mathfrak{C}}(a,b) = \begin{cases} (a,b) & \text{for } a \in A_n,\ b \in B_n \\ c^{\mathfrak{C}} & \text{otherwise,} \end{cases}$$

and where τ is the topology generated by

$$P(S^{\mathfrak{C}}) \,\dot\cup\, \{\{c^{\mathfrak{C}}\} \cup \bigcup_{i \geq n} U_i \mid n \in \omega\},$$

where $P(S^{\mathfrak{C}})$ denotes the power set of $S^{\mathfrak{C}}$, and where $U_i = \emptyset$ for odd i, and where $U_i \subset S^{\mathfrak{C}}$ is a bijection from A_i onto B_i for even i.

Put $T = Th_t((\mathfrak{C},\tau))$. - For $i \in \omega$, let $\varphi_i(x) = Mx \wedge \exists^{\geq i} y\ Rxy$ and put

$$\Phi(x) = \{\varphi_i(x) \mid i \in \omega\}.$$

Thus

(1) $(\mathfrak{C},\tau) \models \Phi[d]$ iff $d \in N$ and d is even.

We show that T and Φ satisfy a and b.

b: Suppose by contradiction, that for some $\psi(x) \in L_t$,

$$T \models_t \exists x\ \psi(x) \quad \text{and} \quad T \models_t \forall x(\psi(x) \rightarrow \varphi_i(x)) \quad \text{for all } i \in \omega.$$

Then, by (1), there is an even number $n \in \mathbb{N}$ such that

(2) $(\mathfrak{C},\tau) \models \psi[n]$.

Let m be the rank of the formula ψ. Then

(3) $((\mathfrak{C},n),\tau) \simeq^t_{m+1} (\mathfrak{C},2m+1),\tau)$.

We leave it to the reader to find a sequence $(I_l)_{l \leq m}$ such that $(I_l)_{l \leq m}$: $((\mathfrak{C},n),\tau) \simeq^t_{m+1} ((\mathfrak{C},2m+1),\tau)$. - (2) and (3) imply $(\mathfrak{C},\tau) \models \psi[2m+1]$. Hence $(\mathfrak{C},\tau) \models \Phi[2m+1]$, which contradicts (1).

a: Suppose $(\mathfrak{C}',\tau')$ is a model of T. We have to show that $(\mathfrak{C}',\tau')$ realizes Φ. For $m \in M^{\mathfrak{C}'}$ put

$$A_m' = \{a \mid a \in C', Q^{\mathfrak{C}'}ma\} \quad \text{and} \quad B_m' = \{b \mid b \in C', R^{\mathfrak{C}'}mb\}.$$

Since

$$(\mathfrak{C},\tau) \models \exists X \ni c\ \forall y \text{ "}X \upharpoonright A_y \times B_y \text{ is a partial injective function from } A_y \text{ into } B_y\text{"}$$

i.e. since $(\mathfrak{C},\tau) \models \chi_1$, where χ_1 is the L_t-sentence

$$\chi_1 = \exists X \ni c\ \forall y\ \forall x_1\ \forall x_2\ \forall z_1\ \forall z_2(Qyx_1 \wedge Qyx_2 \wedge Ryz_1 \wedge Ryz_2 \wedge fx_1z_1 \in X \wedge fx_2z_2 \in X \rightarrow (x_1 = x_2 \leftrightarrow z_1 = z_2)) \quad ,$$

we have

$$(\mathfrak{C}',\tau') \models \chi_1.$$

Similarly

$$(\mathfrak{C},\tau) \models \forall X \ni c\ \exists y \text{ "}X \upharpoonright A_y \times B_y \text{ is a left-total relation"},$$

i.e. $(\mathfrak{C},\tau) \models \chi_2$, where χ_2 is the L_t-sentence

$$\chi_2 = \forall X \ni c\ \exists y\ \forall x(Qyx \rightarrow \exists z(Ryz \wedge fxz \in X)).$$

Hence

$$(\mathfrak{C}',\tau') \models \chi_2.$$

Since $(\mathfrak{C}',\tau') \models \chi_1 \wedge \chi_2$ there is for some $m \in M^{\mathfrak{C}'}$ a total injective function from A_m' to B_m'. Since A_m' is infinite, so is B_m'.

Hence $(\mathfrak{C}',\tau') \models \Phi[m]$.

Let L(I) be the "sublanguage" of L_t introduced at the end of § 7, i.e. L(I) is obtained from $L_{\omega\omega}$ adding the logical symbol I, the formation rule

if φ is a formula, t a term, and x a variable, then $It x\varphi$ is a formula,

and where $Itx\varphi$ is read as "t lies in the interior of the set of x such that φ".

L(I) satisfies an omitting types theorem, which we state in the form

9.2 <u>L(I)-omitting types theorem</u>. Assume L is denumerable. Let T be an L(I)-theory and $\Phi = \{\varphi_n(x) \mid n \in \omega\}$ a set of L(I)-formulas. Suppose that there is no L(I)-formula $\psi(x)$ such that $T \cup \{\exists x \psi(x)\}$ has a topological model and $T \models \forall x(\psi(x) \rightarrow \varphi_n(x))$ for all $n \in \omega$. Then there is a denumerable model $(\mathfrak{A},\sigma)$ of T omitting Φ (i.e. for no $a \in A$, $\mathfrak{A} \models \Phi[a]$ holds).

<u>Proof</u>. Let C be a countable set of new individual constants. As in the proof of the $L_{\omega\omega}$-omitting types theorem, we find an $(L \cup C)(I)$-theory T^*, $T^* \supset T$, having a topological model and satisfying:

(1) Given any $(L \cup C)(I)$ formula $\varphi(x)$ there is a $c \in C$ such that $(\exists x\ \varphi(x) \rightarrow \varphi(c)) \in T^*$.

(2) For any $c \in C$ there is an $n \in \omega$ such that $\neg \varphi_n(c) \in T^*$.

Let $\mathfrak{B}^* = ((\mathfrak{B},(c^B)_{c \in C}),\tau)$ be a model of T. Then $A = \{c^B \mid c \in C\}$ is the universe of a substructure $\mathfrak{A}$ of $\mathfrak{B}$.

The subsets of A of the form

$$\varphi^C = \{c^B \mid \mathfrak{B}^* \models Icx\varphi(x)\},$$

with $\varphi(x) \in (L \cup C)(I)$, are the basis of a topology σ on A.

Put $\mathfrak{A}^* = ((\mathfrak{A},(c^B)_{c \in C}),\sigma)$. - By induction on ψ we show

(+) $\qquad \mathfrak{B}^* \models \psi \quad$ iff $\quad \mathfrak{A}^* \models \psi$.

Then by (2), $(\mathfrak{A},\sigma)$ is a model of T omitting Φ.

To show (+), we only prove the non-trivial cases:

Assume $\mathfrak{B}^* \models \exists x\ \psi(x)$, then by (1), $\mathfrak{B}^* \models \psi(c)$ for some $c \in C$. By induction hypothesis, $\mathfrak{A}^* \models \psi(c)$, hence $\mathfrak{A}^* \models \exists x\ \psi(x)$.

If $\mathfrak{B}^* \models Icx\psi(x)$, then $c^{\mathfrak{B}} \in \psi^C$. Since by induction hypothesis,

$$\psi^C \subset \{a \in A \mid \mathfrak{A}^* \models \psi[a]\} = \psi^{\mathfrak{A}^*},$$

c^B is contained in the σ-interior of $\psi^{\mathfrak{A}^*}$, i.e. $\mathfrak{A}^* \models Icx\psi(x)$.

Now, assume $\mathfrak{A}^* \models Icx\psi(x)$. Then $c^B \in \varphi^C \subset \psi^{\mathfrak{A}^*}$ for some $(L \cup C)(I)$-formula $\varphi(x)$. In particular, $\mathfrak{B}^* \models Icx\psi(x)$. We show that $\mathfrak{B}^* \models \forall y(Iyx\varphi(x) \rightarrow \psi(y))$ (then we obtain $\mathfrak{B}^* \models Icx\psi(x)$ as $\mathfrak{B}^* \models Icx\varphi(x)$). Otherwise $\mathfrak{B}^* \models \exists y(Iyx\varphi(x) \wedge \neg \psi(y))$, i.e. by (1), $\mathfrak{B}^* \models Idx\varphi(x) \wedge \neg \psi(d)$ for some $d \in C$. Hence $d \in \varphi^C$ and $d \notin \psi^{\mathfrak{B}^*}$. By induction hypothesis $d \notin \psi^{\mathfrak{A}^*}$, which contradicts $\varphi^C \subset \psi^{\mathfrak{A}^*}$.

§ 10 $(L_{\omega_1\omega})_t$

This section is devoted to the infinitary language $(L_{\omega_1\omega})_t$. Let L be a similarity type and $\bar{R}$ a denumerable sequence of new relation symbols. Let $(L_{\omega_1\omega})_2$ be the infinitary extension of L_2. Assume $\varphi \in ((L \cup \{\bar{R}\})_{\omega_1\omega})_2$. Denote by $Mod(\exists\bar{R}\varphi)$ the class of topological models of the "Σ_1-sentence over $(L_{\omega_1\omega})_2$" $\exists\bar{R}\varphi$, i.e.

$$Mod(\exists\bar{R}\varphi) = \{(\mathfrak{A},\sigma) \mid (\mathfrak{A},\sigma) \models \exists\bar{R}\varphi\}.$$

Let $Mod^i(\exists\bar{R}\varphi)$ be the class of models of the "invariantization" of $\exists\bar{R}\varphi$, i.e.

$$Mod^i(\exists\bar{R}\varphi) = \{(\mathfrak{B},\tau) \mid (\mathfrak{B},\sigma) \models \exists\bar{R}\varphi \text{ for some basis } \sigma \text{ of } \tau\}.$$

Note that $Mod^i(\exists\bar{R}\varphi) = Mod(\exists\bar{R}\varphi)$ in case φ is invariant for topological structures. In particular, this is true, if $\varphi \in ((L \cup \{\bar{R}\})_{\omega_1\omega})_t$.

Applying methods of Svenonius, Vaught and Makkai (in the form presented in [10]), we show that the denumerable models in $Mod^i(\exists\bar{R}\varphi)$ are the models of a game sentence whose approximations are in $(L_{\omega_1\omega})_t$. Hence, $Mod^i(\exists\bar{R}\varphi)$ is the intersection of $\aleph_1$-many elementary $(L_{\omega_1\omega})_t$-classes. The corresponding covering theorem holds. From this we derive in the usual way the $(L_{\omega_1\omega})_t$-interpolation theorem, and prove that the $(L_{\omega_1\omega})_2$-sentences invariant for topologies are the $(L_{\omega_1\omega})_t$-sentences. We close this section showing how to extend other results from L_t to $(L_{\omega_1\omega})_t$. We remark that Scott's isomorphism theorem does not generalize to $(L_{\omega_1\omega})_t$.

Since our exposition parallels that in [10], it will be helpful if the reader is familiar with this paper. All weak structures in this section are supposed to be models of bas.

Given $(L_{\omega_1\omega})_2$, a countable set C of new constants and a countable set $\mathfrak{U}$ of new set constants, we extend the notion of a Hintikka set Ω from finitary to infinitary logic replacing conditions (ii) and (iii) in 1.4 by its infinitary analogues:

(ii)* If $\bigwedge\Phi \in \Omega$ then $\varphi \in \Omega$ for all $\varphi \in \Phi$.

(iii)* If $\bigvee\Phi \in \Omega$ then $\varphi \in \Omega$ for some $\varphi \in \Phi$.

As above let $\bar{R}$ be a denumerable sequence of relation symbols not in L. Put $L' = L \cup \{\bar{R}\}$. Suppose $\varphi \in (L'_{\omega_1\omega})_2$ is in negation normal form. Take a countable admissible fragment A of $(L'_{\omega_1\omega})_2$ containing φ and L_2. Denote by $A(C,\mathfrak{U})$ the set of sentences obtained from formulas in A by replacing free variables (resp. set variables) by constants in C (resp. in $\mathfrak{U}$).

10.1 For any weak structure $(\mathfrak{B},\tau)$ with denumerable $B \cup \tau$ the following are equivalent:

(a) $(\mathfrak{B},\sigma) \models \exists\bar{R}\varphi$ for some σ with $\tilde{\sigma} = \tilde{\tau}$.

(b) There are a Hintikka set $\Omega \subset A(C,\mathfrak{U})$ with $\varphi \in \Omega$, an onto function $\pi^o : C \to B$, and relations $\pi^1, \pi^2 \subset \mathfrak{U} \times \tau$ satisfying:

(i) if $\chi(v_1,\ldots v_n) \in L_{\omega\omega}$ is atomic or the negation of an atomic formula and $c_1,\ldots,c_n \in C$, then
$$\chi(c_1,\ldots,c_n) \in \Omega \text{ iff } \mathfrak{B} \models \chi[\pi^o(c_1),\ldots,\pi^o(c_n)] .$$

(ii) for any χ of the form $c \in U$: $\chi \in \Omega$ or $\neg\chi \in \Omega$.

(iii) if $U \pi^1 V$ and $c \varepsilon U \in \Omega$, then $\pi^o(c) \in V$

if $U\pi^2V$ and $\pi^o(c) \in V$, then $c\varepsilon U \in \Omega$,

(iv) for any $c \in C$ and $V \in \tau$ with $\pi^o(c) \in V$ there is a $U \in \mathfrak{U}$ such that $c\varepsilon U \in \Omega$ and $U\pi^1V$.

(v) for any $c \in C$ and $U \in \mathfrak{U}$ with $c\varepsilon U \in \Omega$ there is $V \in \tau$ with $\pi^o(c) \in V$ and $U\pi^2V$.

Proof. First suppose that (a) holds, i.e. $(\mathfrak{B},\sigma) \models \exists\bar{R}\varphi$ for some σ with $\tilde{\sigma} = \tilde{\tau}$. We may assume, by a Löwenheim-Skolem argument, that σ is denumerable. Choose interpretations ... of the relation symbols in $\bar{R}$ with $((\mathfrak{B},\ldots),\sigma) \models \varphi$. Take arbitrary onto functions $g: C \to B$ and $h:\mathfrak{U} \to \sigma$ and define Ω to be the set of sentences $\chi(c_1,\ldots,c_n,U_1,\ldots,U_r) \in A(C,\mathfrak{U})$ in negation normal form such that for $\chi(x_1,\ldots,x_n,X_1,\ldots,X_r)$ we have

$$((\mathfrak{B},\ldots),\sigma) \models \chi[g(c_1),\ldots,g(c_n),h(U_1),\ldots,h(U_r)].$$

Then Ω is a Hintikka set containing φ. Since $\tilde{\sigma} = \tilde{\tau}$ we have $(id,\varrho^1,\varrho^2):(\mathfrak{B},\sigma) \simeq^t (\mathfrak{B},\tau)$ for some $\varrho^1,\varrho^2 \subset \sigma \times \tau$ (id being the identity on B). Put

$$\pi^o = g,\ \pi^1 = \varrho^1 \circ h,\ \pi^2 = \varrho^2 \circ h$$

(i.e. $\pi^i = \{(U,V) \mid h(U)\varrho^iV\}$ for $i = 1,2$.)

Then (i) - (v) are fulfilled.

Conversely suppose that Ω, π^o, π^1 and π^2 satisfying (b) are given. Let $((\mathcal{D},\ldots),\sigma)$ be the canonical model of Ω. Then (see 1.4) $D = \{\bar{c} \mid c \in C\}$ and $\sigma = \{\bar{U} \mid U \in \mathfrak{U}\}$. In particular, we have $(\mathcal{D},\sigma) \models \exists\bar{R}\varphi$. It suffices to show that $(\mathcal{D},\sigma) \simeq^t (\mathfrak{B},\tau)$; because then there is a σ' with $\tilde{\sigma}' = \tilde{\tau}$ and $(\mathfrak{B},\sigma') \models \exists\bar{R}\varphi$, hence (a) holds. To show $(\mathcal{D},\sigma) \simeq^t (\mathfrak{B},\tau)$ we define $\varrho = (\varrho^o,\varrho^1,\varrho^2)$ with $\varrho:(\mathcal{D},\sigma) \simeq^t (\mathfrak{B},\tau)$:

For $\bar{c} \in D$ let $\varrho^o(\bar{c}) = \pi^o(c)$ (by (i) and 1.4, ϱ^o is well-defined and is an isomorphism). For $i = 1,2$ set $\varrho^i = \{(\bar{U},V) \mid U\pi^iV\}$. It is easy to show that $\varrho:(\mathcal{D},\sigma) \simeq^t (\mathfrak{B},\tau)$.

Next we express condition (b) in $(\mathfrak{B},\tau)$ by a game sentence Φ, the meaning of a game sentence being defined as usual in terms of an infinite two person game, one played by players $\forall$ and $\exists$. In Φ, c_i and d_i range over C, and U_i and V_i over $\mathfrak{U}$. Φ is defined as

$$\Phi = \forall x_o \bigvee_{c_o} \bigwedge_{d_o} \exists y_o \; \forall X_o \ni x_o \bigvee_{U_o} \bigwedge_{V_o} \bigvee_{k_o \in \{0,1\}} \exists Y_o \ni y_o \bigwedge_{\delta_o \in A(C,\mathfrak{U})} \bigvee_{\theta_o \in A(C,\mathfrak{U})}$$

$$\forall x_1 \bigvee_{c_1} \bigwedge_{d_1} \exists y_1 \; \forall X_1 \ni x_1 \bigvee_{U_1} \bigwedge_{V_1} \bigvee_{k_1 \in \{0,1\}} \exists Y_1 \ni y_1 \bigwedge_{\delta_1 \in A(C,\mathfrak{U})} \bigvee_{\theta_1 \in A(C,\mathfrak{U})}$$

$$\ldots \bigwedge_{n < \omega} N^{c_o d_o U_o V_o k_o \delta_o \theta_o \ldots c_n d_n U_n V_n k_n \delta_n \theta_n}(x_o,\ldots,x_n,y_o,\ldots,y_n,X_o,\ldots,X_n,Y_{i_1},\ldots,Y_{i_r}) .$$

where $Y_{i_1},\ldots,Y_{i_r}$ denote those set variables Y_i such that $k_i = 1$, and where the formulas $N^{c_o \ldots \theta_n}$ are defined as follows:

First let

$$\Theta_n = \{\varphi\} \cup \{c_i \in U_i | i \leq n\} \cup \{d_i \in V_i | i \leq n \text{ and } k_i = 1\}$$
$$\cup \{\neg d_i \in V_i | i \leq n \text{ and } k_i = 0\} \cup \{\theta_o,\ldots,\theta_n\}.$$

There are two cases. Either every one of the conditions (i) - (vi) below is satisfied (case 1) or not (case 2). In case 2, we put $N^{c_o \ldots \theta_n} = \neg x_o = x_o$.

In case 1, we put

$$N^{c_o \ldots \theta_n} = \bigwedge\{\psi | \psi \in L_2 \text{ is an atomic or negated atomic formula,}$$
$$\psi = \psi(x_o,\ldots,x_n,y_o,\ldots,y_n,X_o^+,\ldots,X_n^+, Y_{i_1}^-,\ldots,Y_{i_r}^-) \text{, and}$$
$$\psi(c_o,\ldots,c_n,d_o,\ldots,d_n,U_o,\ldots,U_n,V_{i_1},\ldots,V_{i_r}) \in \Theta_n\} .$$

The conditions are as follows:

(i) For no atomic χ, both χ and $\neg\chi$ belong to Θ_n.

(ii) If $\delta_n \in \Theta_n$ and $\delta_n = \bigvee\Psi$ or $\delta_n = \exists x\psi$ or $\delta_n = \exists X\psi$, then $\theta_n \in \Psi$ or $\theta_n = \psi\frac{c}{x}$ for some $c \in C$, or $\theta_n = \psi\frac{U}{X}$ for some $U \in \mathfrak{U}$.

(iii) If $\delta_n \in \Theta_n$, but

$\delta_n \in \Psi$ for some Ψ such that $\bigwedge\Psi \in \Theta_n$

or $\delta_n = \psi\frac{c}{x}$ for some ψ such that $\forall x\psi \in \Theta_n$

or $\delta_n = \psi\frac{U}{X}$ for some ψ such that $\forall X\psi \in \Theta_n$

or $\delta_n = \psi\frac{c}{x}$ where ψ is atomic and $t = c$, $\psi\frac{t}{x} \in \Theta_n$ for some basic t ,

then $\Theta_n = \delta_n$.

(iv) If $\delta_n = c = c$, then $\Theta_n = \delta_n$.

(v) If neither $\delta_n \in \Theta_n$ nor $\neg\delta_n \in \Theta_n$ and ($\delta_n = \chi(c_1,\ldots,c_s)$ with atomic $\chi(v_1,\ldots,v_s) \in L_{\omega\omega}$) or $\delta_n = c \in U$, then $\Theta_n = \delta_n$ or $\Theta_n = \neg\delta_n$.

(vi) If $\delta_n = t = t$ for some basic t and none of (iii), (iv), (v) is the case, then $\Theta_n = t = c$ for some c.

Now, for any weak model $(\mathfrak{B},\tau)$ of <u>bas</u> with denumerable $B \cup \tau$, we have

<u>10.2</u> $(\mathfrak{B},\tau) \models \Phi$ iff $(\mathfrak{B},\tau)$ satisfies 10.1 (b).

Indeed, assume first that $(\mathfrak{B},\tau)$ satisfies 10.1 (b). Choose Ω, π^o, π^1 and π^2 with the properties listed there. But then using Ω, π^o, π^1, π^2, it is easy to obtain a winning strategy for $\exists$.

Conversely, suppose that $(\mathfrak{B},\tau) \models \Phi$, and let us fix a winning strategy of $\exists$. Let the strategy of $\exists$ be pitted against a play of $\forall$ in which $\forall$ chooses

$$a_n \text{(for } x_n), \; d_n, \; K_n \text{ (for } X_n), \; V_n \text{ and } \delta_n,$$

such that

$$B = \{a_n | n \in \omega\}, C = \{d_n | n \in \omega\},$$

$$\{(a,K) | a \in B, K \in \tau, a \in K\} = \{(a_n,K_n) | n \in \omega\}$$

$$C\times\mathfrak{U} = \{(d_n,V_n) | n \in \omega\}, \; A(C,\mathfrak{U}) = \{\delta_n | n \in \omega\},$$

and such that each element of those sets occurs infinitely often in the corresponding enumeration. Let c_n, b_n (for y_n), U_n, k_n, M_n (for Y_n) and Θ_n be the choices of $\exists$.

Put

$$\Omega = \{\varphi\} \cup \{\Theta_n | n \in \omega\},$$

$$\pi^o = \{(c_n,a_n) | n \in \omega\} \cup \{(d_n,b_n) | n \in \omega\},$$

$$\pi^1 = \{(U_n,K_n) | n \in \omega\} \text{ and } \pi^2 = \{(V_n,M_n) | n \in \omega \text{ and } k_n = 1\}.$$

It is easy to verify that Ω, π^o, π^1, π^2 satisfy the conditions listed in 10.1 (b).

10.1 and 10.2 yield

<u>10.3</u> For any weak model $(\mathfrak{B},\tau)$ of <u>bas</u> with denumerable $B \cup \tau$, we have

$(\mathfrak{B},\tau) \models \Phi$ iff $(\mathfrak{B},\sigma) \models \exists\bar{R}\varphi$ for some σ with $\tilde{\sigma} = \tilde{\tau}$.

We define as usual the approximations of Φ: Given $c_o d_o \ldots k_n \delta_n \Theta_n$ and $i \le n+1$ denote by s_i the sequence $c_o d_o \ldots k_{i-1}\delta_{i-1}\Theta_{i-1}$. By induction on the ordinal α define $\Phi_\alpha^{s_n}$ as follows:

$$\Phi_o^{s_n} = \bigwedge_{1 \le i \le n} N^{s_i} ,$$

$$\Phi_{\alpha+1}^{s_n} = \forall x_n \bigvee_{c_n} \bigwedge_{d_n} \exists y_n \forall X_n \ni x_n \bigvee_{U_n} \bigwedge_{V_n} \bigvee_{k_n \in \{0,1\}} \exists Y_n \ni y_n$$

$$\bigwedge_{\delta_n} \bigvee_{\Theta_n} \Phi_\alpha^{s_n c_n d_n U_n V_n k_n \delta_n \Theta_n} ,$$

$$\Phi_\alpha^{s_n} = \bigwedge_{\beta < \alpha} \Phi_\beta^{s_n} \text{ for a limit } \alpha.$$

Note that for countable α, $\Phi_\alpha^{s_n}$ is an $(L_{\omega_1\omega})_t$-formula. In particular, $\Phi_\alpha^{\emptyset}$ is an $(L_{\omega_1\omega})_t$-sentence. We write Φ_α for $\Phi_\alpha^{\emptyset}$ and call Φ_α the α-th approximation of Φ.

Denote by $\models'$ the consequence relation over denumerable weak models. Vaught has shown (cf. [10]):

10.4 (1) $\models \Phi \rightarrow \Phi_\alpha$ for all α.

(2) $\models' \Phi \leftrightarrow \bigwedge_{\alpha < \aleph_1} \Phi_\alpha$

(3) If ψ is any sentence in $(L^1_{\omega_1\omega})_2$ for some $L^1 \supset L$, then $\models' \Phi \rightarrow \psi$ implies that $\models \Phi_\alpha \rightarrow \psi$ for some $\alpha < \aleph_1$.

We state some consequences of 10.4. Recall that by the definition a topological structure $(\mathfrak{B},\tau)$ is denumerable, if $B \cup \sigma$ is denumerable for some basis σ of τ.

10.5 Theorem. Suppose that $\varphi \in ((L \cup \{R\})_{\omega_1\omega})_2$. Denote by $\mathrm{Mod}^i(\exists\bar{R}\varphi)$ the class of models.

$$\mathrm{Mod}^i(\exists\bar{R}\varphi) = \{(\mathfrak{B},\tau) \mid (\mathfrak{B},\tau) \text{ topological structure and } (\mathfrak{B},\sigma) \models \exists\bar{R}\varphi \text{ for some basis } \sigma \text{ of } \tau\}.$$

Then, over denumerable models, $\mathrm{Mod}^i(\exists\bar{R}\varphi)$ is the intersection of $\aleph_1$-many $(L_{\omega_1\omega})_t$-elementary classes. In particular, if φ itself is an $((L \cup \{\bar{R}\})_{\omega_1\omega})_t$ sentence, then over models $\mathrm{Mod}(\exists\bar{R}\varphi)$ resp. $\mathrm{Mod}(\forall\bar{R}\varphi)$ is the intersection resp. union of $\aleph_1$-many $(L_{\omega_1\omega})_t$-elementary classes.

Proof. Suppose $(\mathfrak{B},\tau)$ is a denumerable topological structure. We have

$(\mathfrak{B},\sigma) \models \exists\bar{R}\varphi$ for some σ with $\tilde{\sigma} = \tau$

iff $(\mathfrak{B},\tau) \models \Phi$ (by 10.3)

iff $(\mathfrak{B},\tau) \models \Phi_\alpha$ for all $\alpha < \aleph_1$ (by 10.4 (2)).

Hence, over denumerable models, $\mathrm{Mod}^i(\exists\bar{R}\varphi) = \bigcap_{\alpha < \aleph_1} \mathrm{Mod}(\Phi_\alpha)$.

10.6 Corollary. (i) Over denumerable models, the class of compact topological spaces is the union of $\aleph_1$-many $(\emptyset_{\omega_1\omega})_t$-elementary classes.
(ii) Over denumerable models, the class of connected topological spaces is the union of $\aleph_1$-many $(\emptyset_{\omega_1\omega})_t$-elementary classes.
(iii) Let $(\mathfrak{B},\tau)$ be a denumerable topological structure. The topological structures that are homeomorphic to $(\mathfrak{B},\tau)$ are the denumerable structures in the intersection of $\aleph_1$-many $(L_{\omega_1\omega})_t$-elementary classes. In particular, any two $(L_{\omega_1\omega})_t$-equivalent denumerable structures are homeomorphic.

Proof. (i) and (ii) are immediate by the preceding. For (iii), let $\varphi \in (L_{\omega_1\omega})_2$ be the "classical" Scott sentence of the two-sorted structure $(\mathfrak{B},\sigma_0)$ for some denumerable basis σ_0 of τ. Then $\mathrm{Mod}^i(\varphi) = \{(\mathfrak{A},\sigma) \mid (\mathfrak{A},\sigma)$ topological structure, $(\mathfrak{A},\sigma) \simeq^t (\mathfrak{B},\tau)\}$.

10.7 Lemma. Assume $\varphi^i \in (L^i_{\omega_1\omega})_2$ for $i = 1,2$, and suppose that φ^2 is invariant for topological structures. Put $L = L^1 \cap L^2$.

If $\models_t \varphi^1 \to \varphi^2$, then $\models_t \varphi^1 \to \psi$ and $\models_t \psi \to \varphi^2$ for some $\psi \in (L_{\omega_1\omega})_t$.

Proof. Let Φ be the game sentence associated with $\exists(R)_{R \in L^1 - L}\varphi^1$ (w.l.o.g. we can assume that $L^1 - L$ only contains relation symbols). By 10.3 and the invariance of φ^2 we have $\models' \Phi \to \varphi^2$. Hence, by parts (1) and (3) of 10.4 we have $\models_t \varphi^1 \to \Phi_\alpha$ and $\models_t \Phi_\alpha^t \to \varphi^2$ for some α.
As a corollary we obtain

10.8 $(L_{\omega_1\omega})_t$ - interpolation theorem. Assume $\varphi^i \in (L^i_{\omega_1\omega})_t$ for $i = 1,2$ and let

$L = L^1 \cap L^2$.

If $\models_t \varphi^1 \to \varphi^2$, then $\models_t \varphi^1 \to \psi$ and $\models_t \psi \to \varphi^2$ for some $\psi \in (L_{\omega_1\omega})_t$.

From 10.7, we get for $\varphi^1 = \varphi^2 = \varphi$ and $L^1 = L^2 = L$:

10.9 Theorem. If $\varphi \in (L_{\omega_1\omega})_2$ is invariant for topologies, then φ is equivalent in topological structures to an $(L_{\omega_1\omega})_t$-sentence.

10.10 Remarks and exercises.

1) The results of this section are true for arbitrary weak structures. In particular, the interpolation theorem holds for $\models$ instead of $\models_t$, and any invariant $(L_{\omega_1\omega})_2$-sentence is equivalent to an $(L_{\omega_1\omega})_t$-sentence. But we should remark, that in case we do not restrict to models of bas, the game sentence associated with a formula $\exists\bar{R}\varphi$ has a more complex prefix.

2) For a topological space (A,σ) denote by A^α the α-th derivate of A:

$$A^0 = A$$

$$A^{\alpha+1} = \text{set of all accumulation points of } A^\alpha$$

$$A^\alpha = \bigcap_{\beta<\alpha} A^\beta \quad \text{for limit } \alpha .$$

Show that for denumerable (A,σ) one has:

(A,σ) is compact iff (i) A^α is finite for some $\alpha < \aleph_1$

(ii) for limit β, any $U \in \sigma$, $A^\beta \subset U$ implies $A^\gamma \subset U$ for some $\gamma < \beta$

(iii) for any α and $U \in \sigma$, if $A^{\alpha+1} \subset U$ then $A^\alpha - U$ is finite.

For any ordinal $\alpha \geq 1$ define φ^α by induction,

$$\varphi^1 = \bigvee_n \exists x_1 \ldots \exists x_n\, \forall U_1 \ni x_1 \ldots \forall U_n \ni x_n\, \forall z(z \in U_1 \vee \ldots \vee z \in U_n),$$

$$\varphi^{\alpha+1} = \bigvee_n \exists x_1 \ldots \exists x_n\, \forall U_1 \ni x_1 \ldots \forall U_n \ni x_n\, \varphi^\alpha_{U_1,\ldots,U_n}$$

where $\varphi^\alpha_{U_1,\ldots,U_n}$ is the formula obtained from φ^α replacing any subformula $\forall x\chi$ by $\forall x(x \in U_1 \vee \ldots \vee x \in U_n \vee \chi)$

$\varphi^{\alpha} = \bigvee_{\beta<\alpha} \varphi^{\beta}$ for limit α.

Note that for $\alpha < \aleph_1$, φ^{α} is an $(L_{\omega_1\omega})_t$-sentence. Show that for a denumerable (A,σ) one has:

(A,σ) is compact and A^{α} is finite iff $(A,\sigma) \models \varphi^{\alpha}$.

Hence, when restricting to denumerable topological spaces, the class of compact spaces is $\bigcup_{\alpha<\aleph_1} \mathrm{Mod}(\varphi^{\alpha})$, i.e. we get a more "concrete" representation of this class as union of $\aleph_1$-many L_t-classes as in 10.6 (i).

3) Let σ_o be a denumerable basis of a denumerable topological structure $(\mathfrak{B},\sigma)$. For any ordinal α, define $\varphi^{\alpha}_{(\mathfrak{B},\sigma_o)}$ extending in the obvious way the definition of $\varphi^{n}_{(\mathfrak{B},\sigma_o)}$ (see §.4). For $\alpha < \aleph_1, \varphi^{\alpha}_{(\mathfrak{B},\sigma_o)}$ is an $(L_{\omega_1\omega})_t$-formula. It is easy to show that the class $\mathfrak{K}$ of topological structures homeomorphic to $(\mathfrak{B},\sigma)$ is the class of denumerable models in $\bigcap_{\alpha<\aleph_1} \mathrm{Mod}(\varphi^{\alpha}_{(\mathfrak{B},\sigma_o)})$.

Thus we get a more "concrete" representation" of $\mathfrak{K}$ as the intersection of $\aleph_1$-many classes than in 10.6 (iii). In particular, as remarked there, any two $(L_{\omega_1\omega})_t$-equivalent denumerable structures are homeomorphic. Note that this is also true for uncountable L: namely, if L is uncountable and $(\mathfrak{B},\tau)$ a denumerable L-structure, then there is a countable $L' \subset L$ such that for each $k \in L - L'$, $k^{\mathfrak{B}}$ is $(L'_{\omega_1\omega})_t$-definable in $\mathfrak{B}\restriction L'$. –

Show that Scott's isomorphism theorem does not generalize to L_t: there is a denumerable structure, which cannot be characterized, within the denumerable structures, up to homeomorphism by an $(L_{\omega_1\omega})_t$-sentence. (Hint: It suffices to find a structure of the form (A,τ,μ) without Scott sentence, where A is countable, τ is a topology on A and μ is a monotone system closed under intersections. Take as (A,τ) the set of rational numbers with its topology and let μ have a basis $\{B_n \mid n \in \omega\}$ with $A = B_o \supset B_1 \supset \ldots$, where $B_i \neq \emptyset$ is perfect and nowhere dense in B_{i-1} and $\bigcap B_i = \emptyset$. Show that for all $\alpha < \aleph_1$, there is a ν such that $(A,\tau,\nu) \not\cong^{t} (A,\tau,\mu)$, but $(A,\tau,\nu) \simeq^{t}_{\alpha} (A,\tau,\mu)$.

4) Note that for $\varphi \in ((L \cup \{\bar{R}\})_{\omega_1\omega})_2$, $\mathrm{Mod}^{i}(\exists\bar{R}\varphi)$ is a $PC_{\omega_1\omega}$-class over $(L_{\omega_1\omega})_t$, if in the definition of $PC_{\omega_1\omega}$-class we allow additional universes (compare 4.5). In particular, if φ has only infinite models, then $\mathrm{Mod}^{i}(\exists\bar{R}\varphi) = \mathrm{Mod}(\exists\bar{S}\psi)$ for some $((L \cup \{\bar{S}\})_{\omega_1\omega})_t$ sentence ψ.

5) Give a "syntactic" proof of the $(L_{\omega_1\omega})_t$-interpolation theorem and of 10.9 in the way indicated in 8.8.4 (extending the axioms and rules to the infinitary language).

6) Prove the effective (= admissible) versions of the above theorems.

Finally, we remark that by the above methods, using the appropriate game sentences, it is possible to generalize the preservation theorems of section 5 to $(L_{\omega_1\omega})_t$. Let us sketch the result for sentences preserved under extensions. Call a sentence $\varphi \in (L_{\omega_1\omega})_t$ in negation normal form existential, if it does not contain any universally quantified individual variable.

Suppose φ is a $((L \cup \{\bar{R}\})_{\omega_1\omega})_t$-sentence. Then for any weak model $(\mathfrak{B},\tau)$ of <u>bas</u> with denumerable $B \cup \tau$, we have

$(\mathfrak{A},\sigma) \models \exists\bar{R}\varphi$ for some $(\mathfrak{A},\sigma)$ with $(\mathfrak{B},\tilde{\tau}) \supset (\mathfrak{A},\tilde{\sigma})$ iff $(\mathfrak{B},\tau) \models \Phi^e$,

where Φ^e is obtained from the game sentence Φ deleting in the prefix all parts $\forall x_n \bigvee_{c_n}$, changing $\forall X_n \ni x_n$ to $\forall X_n \ni y_n$ and changing in the corresponding way the matrix of Φ.

The approximations of Φ^e are existential. In particular, one obtains that an $(L_{\omega_1\omega})_t$-sentence preserved under extensions is equivalent to an existential sentence.

Historical remarks

§ 2 L_t (for $L = \emptyset$) was introduced by T.A. McKee in his papers [12], [13].

§ 3 The notion of an invariant sentence is due to McKee. Compactness, completeness and Löwenheim-Skolem theorem are due to S. Garavaglia [7], [8], [9].

§ 4 For 4.19 see also Garavaglia [8] and P. Bankston [1]. The proof of 4.19 is essentially from [9]. - 4.20 (for $L = \emptyset$) was proved by McKee [13].

§ 5 5.13 is due to Garavaglia [9].

§ 6 In [9] is proved that L_t-equivalence is preserved under direct products with box-topology.

§ 8 A number of results about uniform spaces and proximity spaces are due to J. Strobel [15].

§ 9 The omitting types theorem for L(I) was first proved in Makowsky-Ziegler [11].

§ 10 McKee proved in [13], that countable, $L_{\omega_1\omega}$-elementary equivalent topological spaces are isomorphic.

A number of our theorems are also proved in [9]. Note that some of them were announced in [17].

The main part of the results not credited to other authors in the above are due to the second author. The main part of results due to the first author are in §§ 4,6,10.

References

[1] P. Bankston: Topological Ultraproducts, Ph. D. Thesis, Univ. of Wisconsin (1976)

[2] J. Barwise: Admissible sets and structures, Berlin (1975)

[3] C.C. Chang - H.J. Keisler: Model theory, Amsterdam (1974)

[4] S. Feferman: Persistent and invariant formulas for outer extensions, Comp. Math. 20 (1966), pp. 29-52

[5] S. Feferman - R.L. Vaught: The first-order properties of algebraic systems, Fund. Math. 47 (1959), pp. 57-103

[6] J. Flum: First-order logic and its extensions, in: Logic Conference, Kiel, Lecture Notes in Math. 499, 248-310

[7] S. Garavaglia: Completeness for topological structures, Notices AMS, 75T - E36 (1975)

[8] S. Garavaglia: A topological ultrapower theorem, Notices AMS, 75T - E79 (1975)

[9] S. Garavaglia: Model theory of topological structures, Annals of Math. Logic 14 (1978), pp. 13-37

[10] M. Makkai: Admissible sets and infinitary logic, in: Handbook of mathematical logic, Amsterdam (1977), 233-281

[11] J.A. Makowsky - M. Ziegler: A language for topological structures with an interior operator, Archiv für math. Logik (to appear)

[12] T.A. McKee: Infinitary logic and topological homeomorphisms, Zeitschrift für math. Logik und Grundl. der Math. 21 (1975), 405-408

[13] T.A. McKee: Sentences preserved between equivalent topological bases, Zeitschrift für math. Logik und Grundl. der Math. 22 (1976), 79-84

[14] J.S. Schlipf: Toward model theory through recursive saturation, Journ. of Symb. Logic 43 (1978), 183-206

[15] J. Strobel: Lindström-Sätze in Sprachen für monotone Strukturen. Diplomarbeit, TU Berlin (1978)

[16] S. Williard: General topology, Reading, (1970)

[17] M. Ziegler: A language for topological structures which satisfies a Lindström theorem, Bull. Amer. Math. Soc. 82 (1976), 568-570

§ 1 Topological spaces

In this section we study the expressive power of L_t for topological spaces (A,σ), i.e. for $L = \emptyset$. We want to determine the elementary types of all topological spaces. (Two topological structures are of the same elementary type, if they are L_t-equivalent).

We cannot achieve this aim, if (A,σ) is not a T_3-space. For as we show in part A, the theory of all topological spaces, which satisfy (e.g. only) the separation axiom T_2, is hereditarily undecidable. But a good knowledge of the elementary types of all T_2-spaces should provide a decidability procedure.

In part B we prove that the theory of T_3-spaces is decidable by interpreting countable T_3-spaces in "ω-trees" in such a way that L_t-sentences translate to monadic sentences. Then we use Rabin's result that the monadic theory of ω-trees is decidable (1.24).

The determination of the elementary types of all T_3-spaces is done in part C (1.34;1.41). As an application we get - without using Rabin's result - a decision procedure.

Part D contains two applications of the type analysis in C. We characterize:

1) the T_3-spaces with finitely axiomatizable theory (1.45),

2) the $\aleph_o$-categorical T_3-spaces (1.53).

A. Separation axioms.

We noted in I § 2 that T_2 (= hausdorff) and T_3 (= hausdorff + regular) are expressible by L_t-sentences. Before we will start the study of separation axioms between T_2 and T_3, let us remark that T_o and T_1 also belong to L_t:

$$\forall x\ \forall y(x = y \vee (\exists X \ni x \neg y \in X) \vee (\exists Y \ni y \neg x \in Y))$$
$$\forall x\ \forall y(x = y \vee (\exists X \ni x \neg y \in X))$$

We begin with an example.

1.1 Example. The theory of T_o-spaces is hereditarily undecidable.

Terminology: An L_t-theory T is decidable, if there is an effective procedure which decides whether any given L_t-sentence holds in all topological models of T. T is <u>hereditarily undecidable</u>, if every subtheory $T' \subset \{\varphi \mid T \models_t \varphi\}$ is undecidable. We assume here, L to be finite. In the above example, L is empty.

<u>Proof</u>. We show that the theory of partial orderings is <u>interpretable</u> over the theory of T_o-spaces. That means, that there are L_t-formulas $U(x), \Theta(x,y)$, and for every partial order $(B,\leq)$ there is a T_o-space (A,σ) s.t. $(B,\leq) \simeq (U^{(A,\sigma)}, \Theta^{(A,\sigma)})$. From this and the fact that the theory of partial orderings is hereditarily undecidable, our claim follows as in [18]. - Set

$$U(x) = x = x \quad \text{and} \quad \Theta(x,y) = \forall X \ni x \;\; y \in X.$$

If $(B,\leq)$ is a partial order, set $(A,\sigma) = (B,\sigma_{\leq})$, where $\sigma_{\leq}$ has the basis $\{\{x \mid b \leq x\} \mid b \in B\}$.

1.2 <u>Exercise</u>. a) The class of spaces of the form $(B,\sigma_{\leq})$ can be axiomatized by an L_t-sentence.
b) The L_m-theory of all monotone structures is hereditarily undecidable ($L = \emptyset$).

$\sigma_{\leq}$ is not T_1, if $\leq$ is non-trivial. Thus we have to use another interpretation in the following example (which strengthens 1.1).

1.3 <u>Example</u>. The theory of T_1-spaces is hereditarily undecidable.

<u>Proof</u>. A graph is a set together with a reflexive and symmetric binary relation. It has no isolated points, if every element is related to another one. The theory of graphs without isolated points is hereditarily undecidable [18]. We interpret it over the theory of T_1-spaces by

$$U(x) = \exists y(\neg x = y \wedge \Theta(x,y)) \; , \; \Theta(x,y) = \neg \exists X \ni x \;\; \exists Y \ni y \;\; X \cap Y = \emptyset$$

(In the sequel we tacitly will use abbreviations as $X \cap Y = \emptyset$ for $\forall x(x \in X \rightarrow \neg x \in Y)$). - We leave it to the reader to find enough T_1-spaces which are not T_2, and to complete the proof.

1.3 also follows from the next example.

1.4 <u>Example</u>. The theory of T_2-spaces is hereditarily undecidable.

<u>Proof</u>. We interpret the theory of graphs without isolated points over the

theory of T_2-spaces:

$$U(x) = \exists y(\neg\, x = y \wedge \Theta(x,y))$$

$\Theta(x,y) = \neg\ \exists X \ni x\ \exists Y \ni y\ \bar{X} \cap \bar{Y} = \emptyset$, i.e. "x and y cannot be separated by closed neighborhoods".

In order to construct suitable T_2-spaces we use the following lemma.

1.5 Lemma. a) There is a T_3-space (A',σ') with two decreasing sequences $(U_i)_{i \in \omega}$, $(V_i)_{i \in \omega}$ of open sets satisfying

$$\bar{U}_i \cap \bar{V}_i \neq \emptyset\ ,\ U_i \cap V_i = \emptyset \quad \text{and} \quad \bigcap_{i \in \omega} \bar{U}_i = \bigcap_{i \in \omega} \bar{V}_i = \emptyset\ .$$

b) There is a T_2-space (C,τ) with exactly one pair of distinct points a,b, which are not separable by closed neighborhoods.

Proof. a): Take for A' the Euclidean plane $\mathbb{R}^2$ with its natural topology and set

$$U_i = \{(x,y) | y > 0, x > i\},\ V_i = \{(x,y) | y < 0, x > i\}.$$

b): Let A' be as in a) and set $C = A' \dot{\cup} \{a,b\}$.

$\tau = \{0 \subset C | 0 \cap A' \in \sigma'$, if $a \in 0$ then $U_i \subset 0$ for some i, if $b \in 0$ then $V_i \subset 0$ for some i$\}$.

To prove 1.4, let (B,R) be a graph without isolated points. For every pair $(a,b) \in R$ with $a \neq b$ we choose a T_2-space C_{ab} s.t. a and b is the only pair of distinct points, which is not separable by closed neighborhoods. We can assume that $C_{ab} \cap C_{cd} = \{a,b\} \cap \{c,d\}$. If we put $A = \bigcup_{(a,b) \in R-id} C_{ab}$ and

$$\sigma = \{0 \subset A | 0 \cap C_{ab} \text{ open in } C_{ab} \text{ for all } a,b\},$$

we have

$$(B,R) \simeq (U^{(A,\sigma)}, \Theta^{(A,\sigma)}).$$

A $T_{2,5}$-space is, by definition, a space, where any two distinct points can be separated by closed neighborhoods. In our last example we made use of T_2-spaces, which are not $T_{2,5}$-spaces. The question whether the theory of $T_{2,5}$-spaces is decidable led to the following definition.

1.6 Definition. Let (A,σ) be a topological space.

a) For all ordinals α and subsets B of A the set $\bar{B}^{\alpha}$ is defined by the following recursion:

$$\bar{B}^{0} = B\ ;\ \bar{B}^{\lambda} = \bigcup_{\alpha<\lambda} \bar{B}^{\alpha} \quad \text{if } \lambda \text{ is a limit ordinal;}$$

$$\bar{B}^{\alpha+1} = \{a \mid \bar{B}^{\alpha} \cap \bar{U}^{\alpha} \neq \emptyset \text{ for every neighborhood } U \text{ of } a\}.$$

Finally set $\bar{B}^{\infty} = \bigcup_{\alpha} \bar{B}^{\alpha}$

b) (A,σ) is α-separated (∞-separated), if $\overline{\{a\}}^{\alpha} = \{a\}$ $(\overline{\{a\}}^{\infty} = \{a\})$.

The following properties are easy to check

1) In regular spaces, we have $\bar{B}^{\infty} = \bar{B}$. Thus T_3-spaces are ∞-separated.

2) $\beta \leq \alpha$ implies $\bar{B}^{\beta} \subset \bar{B}^{\alpha}$. In particular (for $\beta \leq \alpha$), α-separated spaces are β-separated.

3) $\overline{B \cup C}^{\alpha} = \bar{B}^{\alpha} \cup \bar{C}^{\alpha}$.

4) (A,σ) is $(\alpha+2)$-separated iff any $a,b \in A$, $a \neq b$, can be separated by α-neighborhoods, i.e. iff there are $U,V \in \sigma$ s.t. $a \in U$, $b \in V$, $\bar{U}^{\alpha} \cap \bar{V}^{\alpha} = \emptyset$.

Thus

T_1 = 1-separated, T_2 = 2-separated, $T_{2,5}$ = 3-separated.

1.7 Remark. Given (A,σ) let τ be the finest regular topology with $\tau \subset \sigma$. Then for $B \subset A$, $\bar{B}^{\infty}$ = τ-closure of B (cf. [26]).

1.8 Lemma. For $\alpha \leq \omega$, α-separatedness is L_t-axiomatizable. (This is not true for $\alpha > \omega$, cf. 1.14).

Proof. Define $\varphi^{0}(X,x) = x \in X$,

$$\varphi^{n+1}(X,x) = \forall Y \ni x \quad \exists y(\varphi^{n}(X,y) \wedge \varphi^{n}(Y,y)).$$

Thus "n-separated" is axiomatized by

$\forall x\, \forall y(\varphi^{n}(\{x\},y) \rightarrow x = y)$ and "ω-separated" by {"n-separated"$\mid n \in \omega$}. "ω-separated" is not finitely axiomatizable (there are n-separated spaces which are not (n+1)-separated, cf. 1.13).

The main result of this part is

1.9 Theorem. The theory of n-separated spaces is heriditarily undecidable for every $n < \omega$.

We will prove this interpreting graphs in n-separated spaces. This cannot be done in ω-separated spaces (cf. 1.14 b).

1.10 Problem. Is the theory of ω-separated spaces decidable?
We start the proof of 1.9 with some definitions.

A system is a pair $\mathfrak{A} = (A,(A_i^n)_{n \in \omega, i \in I})$, where the A_i^n are non-empty subsets of A, $A_i^n \subset A_i^{n+1}$ and where $A_i^n \cap A_{\underline{i}}^n = \emptyset$ implies $A_i^o \cap A_{\underline{i}}^{n+1} = \emptyset$.
As an example take a basis $\{A_i | i \in I\}$, $A_i \neq \emptyset$, of a topology σ on A, and put $A_i^n = \bar{A}_i^n$ (in the sense of 1.6).
$\mathfrak{B} = (B,(B_j^n)_{n \in \omega, j \in J})$ extends $\mathfrak{A} = (A,(A_i^n)_{n \in \omega, i \in I})$, if $A \subset B, I \subset J$, $A_i^n = B_i^n \cap A$ and if $A_i^n \cap A_{\underline{i}}^n = \emptyset$ implies $B_i^n \cap B_{\underline{i}}^n = \emptyset$. - A condition $p = p(x_1,\dots,x_k,\nu_1,\dots,\nu_l)$ is a finite set of expressions of the form

$$x_r \in X_{\nu_s}^n \quad , \quad x_r \notin X_{\nu_s}^n \quad \text{or} \quad X_{\nu_r}^n \cap X_{\nu_s}^m = \emptyset .$$

Given a system $\mathfrak{A} = (A,(A_i^n)_{n \in \omega, i \in I})$, $a_1,\dots,a_k \in A$, $i_1,\dots,i_l \in I$ satisfy the condition p, if all expressions of p are true when interpreting x_s resp. $X_{\nu_s}^n$ by a_s resp. $A_{i_s}^n$. We write $\mathfrak{A} \models p(a_1,\dots,a_k,i_1,\dots,i_l)$.
$\mathfrak{A}$ is generic, if for every system $\mathfrak{B}$ which extends $\mathfrak{A}$, every condition $p(x_1,\dots,x_k,\nu_1,\dots,\nu_l)$, all $a_1,\dots,a_{\underline{k}} \in A$, $i_1,\dots,i_{\underline{l}} \in I$ ($\underline{k} \leq k$, $\underline{l} \leq l$) the following holds: If there are $b_{\underline{k}+1},\dots,b_k \in B$, and $j_{\underline{l}+1},\dots,j_l \in J$ s.t. $\mathfrak{B} \models p(a_1,\dots,b_k,i_1,\dots,j_l)$, then there are $a_{\underline{k}+1},\dots,a_k \in A$, and $i_{\underline{l}+1},\dots,i_l \in I$ s.t. $\mathfrak{A} \models p(a_1,\dots,a_k,i_1,\dots,i_l)$.

1.11 Lemma. Every denumerable system can be extended to a denumerable generic system $\mathfrak{A}$.

Proof. We construct $\mathfrak{A}$ as the "union" of an ascending sequence of denumerable systems. This is a standard procedure to obtain "existentially closed" structures (cf. [21]).(We remark that $\mathfrak{A}$ is uniquely determined.)

1.12 Lemma. Let $\mathfrak{A}$ be generic and let σ be the topology on A with subbasis $\{A_i^o | i \in I\}$. Then

a) (A,σ) is ω-separated.

b) For every n there are two decreasing sequences $(U_i)_{i \in \omega}, (V_i)_{i \in \omega}$ of open subsets satisfying

1) $\mathbb{U}_i^{n+1} \cap \mathbb{V}_i^{n+1} \neq \emptyset$

2) $\mathbb{U}_i^n \cap \mathbb{V}_i^n = \emptyset$

3) $\bigcap_{i \in \omega} \mathbb{U}_i^{n+2} = \bigcap_{i \in \omega} \mathbb{V}_i^{n+2} = \emptyset$

Proof. Let $\mathfrak{A}$ and σ be as in 1.12.

Claim 1. $A_{i_1}^{n+1} \cap \ldots \cap A_{i_k}^{n+1} \cap A_{\underline{i}_1}^{o} \cap \ldots \cap A_{\underline{i}_k}^{o} \neq \emptyset$, and $A_{i_1}^{o} \cap \ldots \cap A_{i_k}^{o} \neq \emptyset$ imply $A_{i_1}^{n} \cap \ldots \cap A_{\underline{i}_k}^{n} \neq \emptyset$.

Proof: Set $B = A \dot{\cup} \{b\}$, $J = I$ and

$$B_i^m = \begin{cases} A_i^m \cup \{b\}, & \text{if } i \in \{i_1,\ldots,\underline{i}_k\},\ m \geq n \\ A_i^m & \text{, otherwise.} \end{cases}$$

$\mathfrak{B}$ is a system. From our assumption follows that $\mathfrak{B}$ extends $\mathfrak{A}$. $b, i_1, \ldots, \underline{i}_k$ satisfy the condition $p(x,\nu_1,\ldots,\underline{\nu}_k) = \{x \in X_{\nu_1}^n, \ldots, x \in X_{\underline{\nu}_k}^n\}$. By genericity of $\mathfrak{A}$, p also is satisfied by $a, i_1, \ldots, \underline{i}_k$ for some $a \in A$. Then $a \in A_{i_1}^n \cap \ldots \cap A_{\underline{i}_k}^n$.

Claim 2: If $a \notin A_{i_1}^{n+1}$, there is an i_2 s.t. $a \in A_{i_2}^{o}$ and $A_{i_1}^n \cap A_{i_2}^n = \emptyset$.

Proof: Choose new elements, and set $B = A \cup \{b_i | i \in I\}$, $J = I \cup \{j\}$, $B_j^m = \{a\} \cup \{b_i | a \in A_i^{m+1}\}$ and

$$B_i^m = \begin{cases} A_i^m \cup \{b_i\}, & \text{if } a \in A_i^{m+1} \\ A_i^m & \text{, otherwise.} \end{cases}$$

$\mathfrak{B}$ is a system, which extends $\mathfrak{A}$. Since $a \in B_j^{o}$ and $B_{i_1}^n \cap B_j^n = \emptyset$, the existence of i_2 follows by genericity.

<u>Claim 3</u>: $A^o_{i_1} \cap \ldots \cap A^o_{i_k} \neq \emptyset$ implies $\overline{A^o_{i_1} \cap \ldots \cap A^o_{i_k}}^n = A^n_{i_1} \cap \ldots \cap A^n_{i_k}$.

This is proved by an easy induction on n using claim 1 and claim 2.

We are now in a position to prove a): Suppose $a \neq a' \in A$. We look at the system $\mathfrak{B}$ defined in the proof of claim 2. Since $a \in B^o_j$, $a' \notin B^n_j$, we have $a \in A^o_i$, $a' \notin A^n_i$ for some $i \in I$. Therefore, $a' \notin \overline{\{a\}}^n$, since $\overline{\{a\}}^n \subset A^n_i$.

This shows $\overline{\{a\}}^n = \{a\}$.

For the proof of b) we fix n.

<u>Claim 4</u>: Suppose $a \in A$, $A^o_{i_1} \cap \ldots \cap A^o_{i_k} \neq \emptyset$,

$A^o_{\underline{i}_1} \cap \ldots \cap A^o_{\underline{i}_k} \neq \emptyset$ and $A^{n+1}_{i_1} \cap \ldots \cap A^{n+1}_{\underline{i}_1} \cap \ldots \cap A^{n+1}_{\underline{i}_k} \neq \emptyset$.

Then there are $i_{k+1}, \underline{i}_{k+1} \in I$ s.t. $A^o_{i_1} \cap \ldots \cap A^o_{i_{k+1}} \neq \emptyset$,

$A^o_{\underline{i}_1} \cap \ldots \cap A^o_{\underline{i}_{k+1}} \neq \emptyset$, $A^{n+1}_{i_1} \cap \ldots \cap A^{n+1}_{i_{k+1}} \cap A^{n+1}_{\underline{i}_1} \cap \ldots \cap A^{n+1}_{\underline{i}_{k+1}} \neq \emptyset$,

$A^n_{i_{k+1}} \cap A^n_{\underline{i}_{k+1}} = \emptyset$, $a \notin A^{n+2}_{i_{k+1}}$ and $a \notin A^{n+2}_{\underline{i}_{k+1}}$.

<u>Proof</u>: We define an extension $\mathfrak{B}$ of $\mathfrak{A}$ by

$$B = A \,\dot{\cup}\, \{b,\underline{b},c\} \;, \quad J = I \,\dot{\cup}\, \{j,\underline{j}\},$$

and for $i \in I$

$$B^m_i = A^m_i \cup C^m_i$$

where $C^m_i \subset \{b,\underline{b},c\}$ and

$c \in C^m_i$ iff $m > n$ and $i \in \{i_1,\ldots,i_k,\underline{i}_1,\ldots,\underline{i}_k\}$

$b \in C^m_i$ iff $i \in \{i_1,\ldots,i_k\}$,

$\underline{b} \in C^m_i$ iff $i \in (\underline{i}_1,\ldots,\underline{i}_k\}$.

$$B^m_j = \begin{cases} \{b\} & \text{if } m \leq n \\ \{b,c\} & \text{if } m > n \end{cases} ,$$

$$B^m_{\underline{j}} = \begin{cases} \{\underline{b}\} & \text{if } m \leq n \\ \{\underline{b},c\} & \text{if } m > n \end{cases} .$$

We have $b \in B^o_{i_1} \wedge \ldots \wedge B^{o}_{i_k} \wedge B^o_j$, $\underline{b} \in B^o_{\underline{i}_1} \wedge \ldots \wedge B^o_{\underline{i}_k} \wedge B^o_{\underline{j}}$,

$c \in B^{n+1}_{i_1} \wedge \ldots \wedge B^{n+1}_{i_k} \wedge B^{n+1}_j \wedge B^{n+1}_{\underline{i}_1} \wedge \ldots \wedge B^{n+1}_{\underline{j}}$, $B^n_j \wedge B^n_{\underline{j}} = \emptyset$,

$a \notin B^{n+2}_j$ and $a \notin B^{n+2}_{\underline{j}}$. Our claim follows from the genericity of $\mathfrak{A}$.

To prove b) we choose an enumeration $\{a_k\}_{k \in \omega}$ of A. Using claim 4 we construct two sequences $i_1, i_2, \ldots; \underline{i}_1, \underline{i}_2, \ldots$ of elements of I s.t.

$A^o_{i_1} \cap \ldots \cap A^o_{i_k} \neq \emptyset$, $A^o_{\underline{i}_1} \cap \ldots \cap A^o_{\underline{i}_k} \neq \emptyset$, $A^{n+1}_{i_1} \cap \ldots \cap A^{n+1}_{i_k} \cap A^{n+1}_{\underline{i}_1} \cap \ldots \cap A^{n+1}_{\underline{i}_k} \neq \emptyset$,

$A^n_{i_k} \cap A^n_{\underline{i}_k} = \emptyset$, $a_k \notin A^{n+2}_{i_k}$ and $a_k \notin A^{n+2}_{\underline{i}_k}$. - Set $U_k = A^o_{i_1} \cap \ldots \cap A^o_{i_{k+1}}$

and $V_k = A^o_{\underline{i}_1} \cap \ldots \cap A^o_{\underline{i}_{k+1}}$.

1.13 Lemma. There is an (n+2)-separated space (C, τ) with exactly one pair of distinct points a,b, which are not separable by (n+1)-neighborhoods.

Proof. Let (A, σ) be as in 1.12 and set $C = A \mathbin{\dot{\cup}} \{a,b\}$,

$\tau = \{O \subset C \mid O \cap A \in \sigma$, if $a \in O$ then $U_i \subset O$ for some $i \in \omega$,
if $b \in O$ then $V_i \subset O$ for some $i \in \omega\}$.

The lemma follows from 1) – 3) below which are proved by induction on $m \leq n+1$.

1) $\bar{U}^m$ (w.r.t.τ) = $\bar{U}^m$ (w.r.t.σ) for all sufficiently small neighborhoods U of any $c \in A$.

2) $\overline{\{a\} \cup U_i}^m = \{a\} \cup \bar{U}^m_i$ (w.r.t.σ)

3) $\overline{\{b\} \cup V_i}^m = \{b\} \cup \bar{V}^m_i$ (w.r.t.σ)

Proof of theorem 1.9: We interpret the theory of graphs without isolated points over the theory of (n+2)-separated spaces using the formulas

$$U(x) = \exists y(\neg\, x = y \wedge \Theta(x,y))$$

$\Theta(x,y)$ = "x and y are not separable by (n+1)-neighborhoods" (see 1.8).

We now proceed as in the proof of 1.4, where we used 1.5 b) instead of 1.13. It is helpful to prove by induction on $m \leq n+1$ that for any $c \in A$ and all sufficiently small neighborhoods U of c,

$$\overline{U}^m(\text{w.r.t.}\sigma) = \bigcup_{(a,b)\in R} \overline{U \cap C_{ab}}^{\,m} \quad (\text{w.r.t.}\tau_{ab}) .$$

We return to our problem 1.10. The following theorem shows that it may be hard to prove the undecidability of the theory of ω-separated spaces.

1.14 Theorem. a) A topological space is ω-separated iff it is L_t-equivalent to a space where every two points can be separated by clopen neighborhoods (and which is therefore ∞-separated).

In fact every denumerable recursively saturated ω-separated space has this property. (As already indicated in I. § 4, we call a topological structure $(\mathfrak{A},\sigma)$ recursively saturated, if for some basis β of σ the two-sorted structure $(\mathfrak{A},\beta)$ is recursively saturated.)

b) With respect to the theory of ω-separated spaces every L_t-formula $\varphi(x_1,\ldots,x_n)$ is equivalent to a boolean combination of formulas of the form $x_i = x_j$, $\psi(x_i)$.

Proof. a) It is easily shown by induction that $\overline{U}^\alpha = U$ for any clopen U and every α. Whence spaces where any two points can be separated by clopen neighborhoods are ∞-separated.

Every space is L_t-equivalent to a denumerable recursively saturated space (A,σ). Suppose that (A,σ) is ω-separated. We show that (A,σ) is ∞-separated. Let β be a basis of σ, for which (A,β) is a recursively saturated two-sorted structure. If $U \in \beta$ and $a \notin \overline{U}^\omega$, there is, for every n, a neighborhood $V \in \beta$ of a s.t. $\overline{V}^n \cap \overline{U}^n = \emptyset$. By saturatedness, there is $V \in \beta$ which contains a and satisfies $\overline{V}^n \cap \overline{U}^n = \emptyset$ for all $n \in \omega$, i.e. $\overline{V}^\omega \cap \overline{U}^\omega = \emptyset$. This shows $\overline{U}^\omega = \overline{U}^{\omega+1}$. The proof of $\{a\} = \overline{\{a\}}^\omega = \overline{\{a\}}^{\omega+1}$ (for any $a \in A$) is similar. From this follows by induction that $\overline{U}^\alpha = \overline{U}^\omega$ and $\{a\} = \overline{\{a\}}^\omega = \overline{\{a\}}^\alpha$ holds for all $\alpha \geq \omega$. - Let $p,q \in A$, $p \neq q$. To separate p and q by a clopen set, we only will use that $\overline{\{p\}}^\infty \cap \overline{\{q\}}^\infty = \emptyset$, and that $A = \{a_i \mid i \in \omega\}$ is denumerable.

Set $P_o = \{p\}$, $Q_o = \{q\}$ and suppose that P_i, Q_i with $\overline{P}_i^\infty \cap \overline{Q}_i^\infty = \emptyset$ have already been defined.

Case 1. $a_i \notin \overline{Q}_i^\infty$. Then, for all α, there is a $V_\alpha \in \sigma$ s.t. $a_i \in V_\alpha$, $\overline{V}_\alpha^\alpha \cap \overline{Q}_i^\alpha = \emptyset$. Choose V s.t. $V = V_\alpha$ for arbitrarily large α. Then

$\overline{V}^{\infty} \cap \overline{Q}_i^{\infty} = \emptyset$. Set $P_{i+1} = P_i \cup V$ and $Q_{i+1} = Q_i$.

Case 2: $a_i \in \overline{Q}_i^{\infty}$. Then $a_i \notin \overline{P}_i^{\infty}$ and as in the first case, we find $a_i \in V \in \beta$ s.t. $\overline{V}^{\infty} \cap \overline{P}_i^{\infty} = \emptyset$ and set $P_{i+1} = P_i$ and $Q_{i+1} = Q_i \cup V$.

$\bigcup_{i \in \omega} P_i$ is a clopen set which separates p and q.

1.15 Exercise. Let $(\mathfrak{A},\sigma)$ be a topological structure and $B,C \subset A$. Then, $\overline{B}^{\omega} \cap \overline{C}^{\omega} = \emptyset$ iff there is a topological structure $((\mathfrak{A}',B',C'),\sigma')$ such that $((\mathfrak{A}',B',C'),\sigma') \equiv^t ((\mathfrak{A},B,C),\sigma)$ and B' and C' can be separated by a clopen set.

Proof of 1.14 b). A standard compactness argument shows that it suffices to prove:

> Any two n-tuples $a_1,\ldots,a_n \in A$, $b_1,\ldots,b_n \in B$ satisfying the same formulas of the form $x_i = x_j$, $\psi(x_i)$ in ω-separated spaces (A,σ) and (B,τ) satisfy the same formulas $\varphi(x_1,\ldots,x_n)$.

We can assume that for some bases α of σ and β of τ, the pair $((A,\alpha),(B,\beta))$ is a denumerable recursively saturated weak structure. Then (A,α) and (B,β) are homeomorphic and will be identified, $(A,\alpha) = (B,\beta)$. We may assume that $a_1,\ldots,a_n$ are distinct. Then $b_1,\ldots,b_n$ must be distinct too. Whether $a_i = b_j$ or not causes a lot of cases.

We treat only a typical example:

$n = 6$

$a_2 = b_1,\ a_3 = b_2,\ a_1 = b_3,\ a_5 = b_4,\ a_i \neq b_j$ otherwise.

By a) we find disjoint clopen sets U_i, U', U'' s.t. $a_i \in U_i$, $b_5 \in U'$ and $b_6 \in U''$. Since a_i and b_i satisfy the same L_t-formulas, we find an automorphism f_i of (A,σ) which maps a_i on b_i.

We set

$$V_1 = U_1 \cap f_1^{-1}(U_2 \cap f_2^{-1}(U_3)),$$

$$V_2 = f_1(V_1)\ ,\quad V_3 = f_2(V_2),$$

$$V_4 = U_4 \cap f_4^{-1}(U_5 \cap f_5^{-1}(U'))$$

$$V_5 = f_4(V_4),\quad V' = f_5(V_5),$$
$$V_6 = U_6 \cap f_6^{-1}(U''),\quad V'' = f_6(V_6)\ .$$

$V_1,\ldots,V_6,V,V'$ are disjoint neighborhoods of $a_1,\ldots,a_6,b_5,b_6$. The union of the functions

$$f_1 \restriction V_1,\ f_2 \restriction V_2,\ f_1^{-1}f_2^{-1} \restriction V_3,\ f_4 \restriction V_4,\ f_5 \restriction V_5,\ f_4^{-1}f_5^{-1} \restriction V',\ f_6 \restriction V_6,$$
$$f_6^{-1} \restriction V'',\quad id \restriction A\setminus(V_1\cup\ldots\cup V_6 \cup V' \cup V'')$$

is an automorphism of (A,σ) mapping $a_1,\ldots,a_6$ onto $b_1,\ldots,b_6$. Therefore, these two 6-tuples satisfy the same formulas $\varphi(x_1,\ldots,x_6)$.

1.16 Exercises. a) Every uniform structure is L_m^2-equivalent to a uniform structure $(\mathfrak{A},\mu)$, where any two points can be separated by a uniform open set B (i.e. there is $N \in \mu$ s.t. $a \in B$ implies $N(a) \subset B$).

b) Prove the result corresponding to 1.14 b) for uniform spaces.

c) Show that in proximity spaces the relation

$$x \sim y \quad \text{iff} \quad \exists(X,Y)\ (x \in X \wedge y \in Y \wedge \forall z(z \in X \vee z \in Y)$$

may be non-trivial. (It is open whether the thery of proximity spaces is decidable).

B The decidability of the theory of T_3-spaces.

In a certain sense T_3 is the strongest separation axiom which is expressible in L_t: A T_0-topology is called 0-dimensional, if it has a basis of clopen sets. We have

1.17 Theorem. A topological structure is T_3 iff it is L_t-equivalent to a 0-dimensional topological structure. - In any denumerable T_3-structure disjoint closed sets can be separated by clopen sets.

Proof. In T_3-spaces, we have $\bar{B}^{\infty} = \bar{B}$. In particular, if P and Q are closed and disjoint, then $\bar{P}^{\infty} \cap \bar{Q}^{\infty} = \emptyset$. The proof of 1.14 shows that P and Q can be separated by a clopen set, if the universe is countable.

0-dimensional spaces are T_3. Thus every space L_t-equivalent to a 0-dimensional space is T_3.

Now let $(\mathfrak{A},\sigma)$ be a T_3-structure. If L is denumerable, we find a denumerable T_3-structure $(\mathfrak{B},\tau)$ L_t-equivalent to $(\mathfrak{A},\sigma)$. By our first remark, τ is 0-dimensional. - If L is uncountable, let $(\mathfrak{B},\tau)$ be ω_1-saturated and L_t-equivalent to $(\mathfrak{A},\sigma)$. We want to show that τ is 0-dimensional. Let U be an open neighborhood of $b \in B$. By regularity there exists a sequence $U \supset U_0 \supset U_1 \supset \dots$ of open neighborhoods of b s.t. $\bar{U}_{i+1} \subset U_i$. $\bigcap_{i \in \omega} U_i = \bigcap_{i \in \omega} \bar{U}_i$ is closed, and - by the next lemma - open.

1.18 <u>Lemma</u>. Let (B,τ) be ω_1-saturated. Then τ is closed under countable intersections.

<u>Proof</u>. Choose a basis β s.t. the two-sorted structure (B,β) is ω_1-saturated. Suppose $O_i \in \tau$ and $b \in \bigcap_{i \in \omega} O_i$. Choose neighborhoods $V_i \in \beta$ with $b \in V_i$, $V_i \subset O_i$. The type

$$\{c_o \in X\} \cup \{\forall x(x \in X \rightarrow x \in C_i) | i \in \omega\}$$

is finitely satisfiable in $((B,b,V_o,V_1,\dots),\beta)$.
Whence there is a $V \in \beta$ with $b \in V$ and $V \subset V_i \subset O_i$ for $i = 0,1,2,\dots$.
Therefore, $\bigcap_{i \in \omega} O_i$ is open.

1.19 <u>Exercise</u>. Give a finite axiomatization of the class of all topological spaces which are L_t-equivalent to a space with a basis α s.t. $C,D \in \alpha$ implies $C \subset D$ or $D \subset C$ or $C \cap D = \emptyset$.

By the Löwenheim-Skolem theorem it is enough to know the elementary properties of all denumerable spaces. We use the following presentation of the denumerable T_3-spaces.

1.20 <u>Definition</u>. An <u>ω-tree</u> is a denumerable partial ordering $(T,\leq)$, where all sets $\{b | b \leq a\}$, $a \in T$, are finite and linearly ordered by $\leq$.

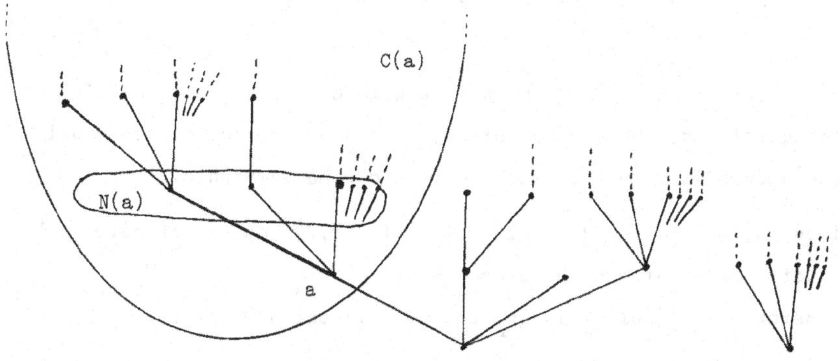

We will use the notations:

$C(a) = \{b \mid a \leq b\}$, the "cone" of a,

$N(a) = \{b \mid \forall x(x \leq a \leftrightarrow x < b)\}$, the immediate successors of b.

We define a topology $\tau_{\leq}$ on T using the sets

$$U_{\Delta}(a) = \{a\} \cup U\{C(b) \mid b \in N(a) - \Delta\} ,$$

where Δ is a finite subset of $N(a)$, as a basis of the neighborhoods of a. $\tau_{\leq}$ is a T_3-topology with the denumerable basis $\{U_{\Delta}(a) \mid a \in T, \Delta \text{ finite}\}$ of clopen sets.

1.21 Theorem. Every denumerable T_3-space is homeomorphic to a space of the form $(T,\tau_{\leq})$, where $(T,\leq)$ is an ω-tree.

Proof. Let (T,τ) be a denumerable T_3-space. We want to find an ω-tree structure $\leq$ on T s.t. $\tau = \tau_{\leq}$. First fix an enumeration of T. Then we construct a set $\mathfrak{T}$ of non-empty clopen subsets of T s.t.

i) $(\mathfrak{T},\supset)$ is an ω-tree with smallest element T,

ii) if a_A is the first element of $A \in \mathfrak{T}$, then there are clopen sets $A = B_0 \supset B_1 \supset \ldots$ forming a basis of the neighborhoods of a_A s.t. the immediate successors of A are just those differences $B_i - B_{i+1}$, which are not empty.

$A \mapsto a_A$ yields a bijection from $\mathfrak{T}$ onto T. Define $\leq$ by

$$a_A \leq a_B \quad \text{iff} \quad A \supset B$$

Then $\tau = \tau_{\leq}$. Note that A is the cone of a_A.

1.22 Corollary. All denumerable T_3-spaces without isolated points are homeomorphic to $(\mathbb{Q},\tau_{<})$, the topological space of the rationals with the order topology.

Proof. The construction in 1.21 yields a tree with a smallest element. In case the topological space has no isolated points, every point has countably many immediate successors. But all such ω-trees are homeomorphic.

1.23 Exercises. a) Show: $(T,\tau_{\leq})$ is compact iff $(T,\leq)$ has no infinite path and only a finite number of minimal elements.

b) Every denumerable T_3-topology is induced by a linear ordering.

c) Deduce 1.24 below from b) and the decidability of the theory of linear orderings. (cf. [13]).

1.24 Theorem. The theory of T_3-spaces is decidable.

Proof. By [17] the weak second order theory T_w of all ω-trees is decidable, i.e.

$$T_w = Th_{L_2'}\{((T,\leq),P_\omega(T)) \mid (T,\leq)\ \omega\text{-tree}\}$$

is decidable. ($P_\omega(T)$ denotes the set of finite subsets of T). - We assign to every L_t-sentence φ an L_2'-sentence $\hat{\varphi}$ ($L = \emptyset$, $L' = \{\leq\}$) s.t.

$$(T,\tau_{\leq}) \models \varphi \quad \text{iff} \quad ((T,\leq),P_\omega(T)) \models \hat{\varphi}\ .$$

To obtain $\hat{\varphi}$ we replace in φ the set quantifiers as indicated by ($Q = \exists,\forall$)

$$\ldots QX \ni t \ldots s \in X \ldots \quad \mapsto \ldots QX \ldots U_X(t) \ldots \quad ,$$

where $U_X(t)$ is

$$(t \leq s \ \wedge \ \forall x(x \in X \ \wedge \ x \leq s \ \rightarrow \ x \leq t))\ .$$

Then, we have by 1.21 and the Löwenheim-Skolem theorem

$$\varphi \text{ holds in all } T_3\text{-spaces iff } \hat{\varphi} \in T_w .$$

1.25 Corollary. The theory of T_3-spaces with unary relations is decidable.

Proof. The weak second order theory of ω-trees with unary relations is decidable [17].

1.26 Exercises. a) It is well known that a formula $p(P_1,\ldots,P_n)$ of propositional calculus is intuitionistically valid iff in every T_3-space p is satisfied by all sequences $O_1,\ldots,O_n$ of open sets (where the connectives are interpreted in the Heyting-algebra of open sets). Show that the set of intuionistically valid formulas is decidable.
b) Show that the theory of regular spaces is decidable.

1.25 corresponds in classical model theory to the decidability of the theory of universes with unary relations, which is easy to prove. The classical result that the theory of universes with a unary functions is decidable (cf. [17]) has only the following negative counterpart.

1.27 Remark. The L_t-theory of all two-sorted topological structures $((A,\sigma),(B,\tau),f)$, where σ and τ are T_3 and $f:A \to B$ is continuous, open and surjective, is hereditarily undecidable.

Proof. We interprete the theory of graphs without isolated points over the theory in discussion. The formulas are

$$U(x) = \exists y(\neg\, y = x \wedge \Theta(x,y)),$$

$$\Theta(x,y) = \forall X \ni x\ \forall Y \ni y\ \exists x_1\ \exists x_2\ \exists y_1\ \exists y_2(\neg\, x_1 = x_2 \wedge \neg\, y_1 = y_2 \wedge y_1 \in Y$$

$$\wedge\, y_2 \in Y \wedge x_1 \in X \wedge x_2 \in X \wedge f(x_1) = f(x_2) = f(y_1) = f(y_2)).$$

If (C,R) has no isolated points and, say, is denumerable, choose an enumeration $((a_i,b_i) \mid i \in \omega)$ of R, where every pair occurs infinitely many times. Take new elements c_b^i, d_b^i for $b \in C$, $i \in \omega$, s.t. $c_a^i = d_a^i$ for $a \notin \{a_i,b_i\}$ are the only equalities.

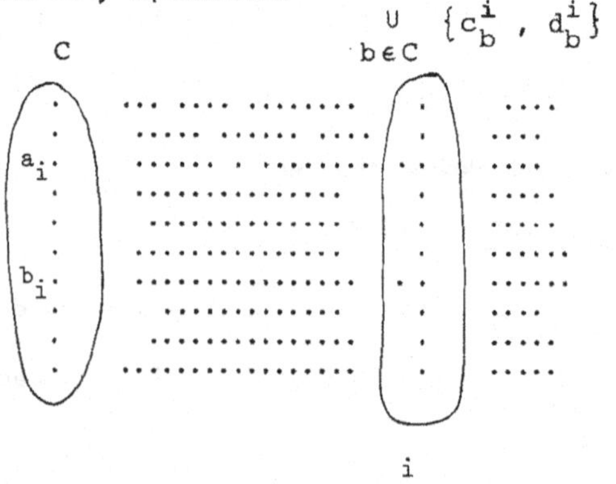

Set $A = C \cup \{c_b^i \mid b \in C, i \in \omega\} \cup \{d_b^i \mid b \in C, i \in \omega\}$. Let σ be the topology, where all c_b^i, d_b^i are isolated, and the sets $\{b\} \cup \Delta$, where $b \in C$ and where Δ is cofinite in $\{c_b^i \mid i \in \omega\} \cup \{d_b^i \mid i \in \omega\}$, form a basis of the neighborhoods of b. Let B be the partition $\{C\} \cup \{\bigcup_{b \in C} \{c_b^i, d_b^i\} \mid i \in \omega\}$, f the projection and τ the quotient topology. Then $U^{\mathfrak{A}} = C$ and $\Theta^{\mathfrak{A}} = R$.

Note that a continuous map f yields an equivalence relation $\equiv$ with closed classes: $x \equiv y$ iff $f(x) = f(y)$. Hence, the theory of all T_3-topological structures $((A,\equiv),\sigma)$, where $\equiv$ is an equivalence relation with closed classes is hereditarily undecidable.

We conclude part B with another application of our tree method.

1.28 <u>Theorem</u>. The theory of hausdorff uniform spaces is decidable. (We mean the L^2_m-theory of hausdorff uniform spaces, see p. 53).

<u>Proof</u>. We call a non-empty subset A of an ω-tree $(T,\leq)$ <u>good</u>, if every point of T has exactly one immediate successor which does not belong to A. Then, for every $a \in A$, there is a unique path X_a (= maximal linearly ordered subset) with

$$A \cap C(a) \cap X_a = \{a\}.$$

The equivalence relations U_n,

$$U_n = \{(a,b) \mid |X_a \cap X_b| \geq n\}$$

for $n \in \omega$, form a basis of a hausdorff uniformity $\nu_{(T,\leq,A)}$ on A.

1.29 <u>Lemma</u>. Every hausdorff uniform space is L^2_m-equivalent to a uniform space of the form $(A,\nu_{(T,\leq,A)})$, where A is a good subset of some ω-tree T.

<u>Proof</u>. Let (B,ν) be an arbitrary hausdorff uniform space. Take a weak structure (B',β') L^2_2-equivalent to (B,ν) and ω_1-saturated. β' is the basis of a uniformity ν', which is closed under countable intersections. ν' also has a basis β'' consisting of equivalence relations. For, let $N_o \in \nu'$, then there is a sequence $N_o \supset N_1 \supset N_2 \supset \ldots$ of elements of ν' s.t. . $N_{i+1} \circ N^{-1}_{i+1} \subset N_i$. $\bigcap_{i \in \omega} N_i \in \nu'$ is an equivalence relation.

Now choose a denumerable weak structure (A,α) with $(A,\alpha) \equiv_{L^2_2} (B',\beta'')$. If μ is the monotone system generated by α, we have $(A,\mu) \equiv_{L^2_m} (B,\nu)$. μ has a descending basis $A^2 = U_o \supset U_1 \supset \ldots$ of equivalence relations. Since μ is hausdorff, we find a sequence $A_o \subset A_1 \subset \ldots$ s.t. $A = \bigcup_{i \in \omega} A_i$ and A_i is a transversal set for U_i. The pairs $({}^a/_{U_i},i)$ with the ordering defined by

$$({}^a/_{U_i},i) \leq ({}^b/_{U_j},j) \quad \text{iff} \quad i \leq j \text{ and } b \in {}^a/_{U_i}$$

form an ω-tree $(T,\leq)$. If we identify $a \in A$ with $({}^a/_{U_i},i)$, where $a \in A_i - A_{i-1}$, A becomes a good subset of $(T,\leq)$ and $\nu_{(T,\leq,A)} = \mu$.

We return to the proof of 1.28. It suffices to decide, if an L^2_m-sentence holds in all $(A,\nu_{(T,\leq,A)})$.

A subset X of the ω-tree T is <u>bounded</u>, if the set of numbers $h(x) = |\{t \mid t \leq x\}|$ with $x \in X$ are bounded. Let $P_b(T)$ be the set of all bounded subsets of T. In [22] it is shown that

$$Th_{L_2'}(\{((T,\leq,A),P_b(T)) \mid (T,\leq)\ \omega\text{-tree},\ A \subset T\})$$

is decidable ($L' = [\leq,A]$). 1.28 is proved, if

(i) "A is good" is expressible in $((T,\leq,A),P_b(T))$ by an L_2'-sentence.

(ii) for every L^2_m-sentence φ we can effectively find an L_2'-sentence $\hat{\varphi}$ s.t. for any ω-tree $(T,\leq)$ and any good subset A of T, we have

$$(A,\nu_{(T,\leq,A)}) \models \varphi \quad \text{iff} \quad ((T,\leq,A),P_b(T)) \models \hat{\varphi}.$$

(i) is clear. – (ii): Fix a good $A \subset T$. For $S \subset T$ set $U_S = \{(a,b) \mid X_a \cap X_b \not\subset S\}$. Since $t \in X_a$ is expressible by

$$t \leq a \vee (a < t \wedge (\forall x(x \leq t \wedge a < x) \rightarrow \neg Ax)),$$

and $(a,b) \in U_S$ by

$$\exists x(\neg x \in S \wedge x \in X_a \wedge x \in X_b),$$

there is for every L^2_2-sentence φ an L_2'-sentence $\hat{\varphi}$ s.t.

$$(A,\{U_S \mid S \in P_b(T)\}) \models \varphi \quad \text{iff} \quad ((T,\leq,A),P_b(T)) \models \hat{\varphi}.$$

But the U_S, $S \in P_b(T)$, form a basis of $\nu_{(T,\leq,A)}$.

Whence

$$(A,\nu_{(T,\leq,A)}) \models \varphi \quad \text{iff} \quad ((T,\leq,A),P_b(T)) \models \hat{\varphi}.$$

for all L^2_m-sentences φ .

1.30 <u>Exercises</u>. a) Prove the decidability of the theory of uniform spaces.
b) The theory of structures of the form (A,σ,τ), where σ and τ are T_3-topologies, is hereditarily undecidable.

C. The elementary types of T_3-spaces

Let (A,σ) be a topological space. We partition A in classes of points of the same "n-type". All points have the same 0-type. a and b are of the same (n+1)-type, if they are accumulation points of the same n-types. More precisely,

1.31 <u>Definition.</u> We define the set $\mathfrak{T}_n$ of n-types by induction,

$$\mathfrak{T}_0 = \{*\} \quad \text{and} \quad \mathfrak{T}_{n+1} = P(\mathfrak{T}_n).$$

Let $\mathfrak{T} = \bigcup_{n \in \omega} \mathfrak{T}_n$. - The n-type of $a \in A, t_n(a)$, is defined inductively by

$$t_0(a) = * ,$$

$$t_{n+1}(a) = \{\alpha \in \mathfrak{T}_n \mid \text{in every neighborhood of } a \text{ there is } b \neq a \text{ with } t_n(b) = \alpha\}.$$

Sometimes we write $t_n(A,a)$ or $t_n((A,\sigma),a)$ to stress the dependence of $t_n(a)$ upon (A,σ).

There are two 1-types, $\emptyset$ and $\{*\}$. A point has 1-type $\emptyset$ iff it is isolated.- A point a has 2-type $\emptyset$ iff a is isolated. If a is not isolated, a has 2-type $\{\{*\}\}$ or $\{\emptyset\}$ or $\{\{*\},\emptyset\}$ according as a is only an accumulation point of the set of accumulation points or only of the set of isolated points or of both sets.

For $m \leq n$, the n-type of a determines the m-type of a: If we define for $\alpha \in \mathfrak{T}$ and $s \in \omega$ the s-type $(\alpha)_s$ by $(\alpha)_0 = *$ and $(\alpha)_{s+1} = \{(\beta)_s \mid \beta \in \alpha\}$, we can prove, for $m \leq n$, that $(t_n(a))_m = t_m(a)$.

This gives rise to a tree structure $\leq$ on the "disjoint union" of the $\mathfrak{T}_n$:

$$\alpha \leq \beta \quad \text{iff} \quad \alpha \in \mathfrak{T}_m,\ \beta \in \mathfrak{T}_n,\ m \leq n \ \text{and}\ \alpha = (\beta)_m.$$

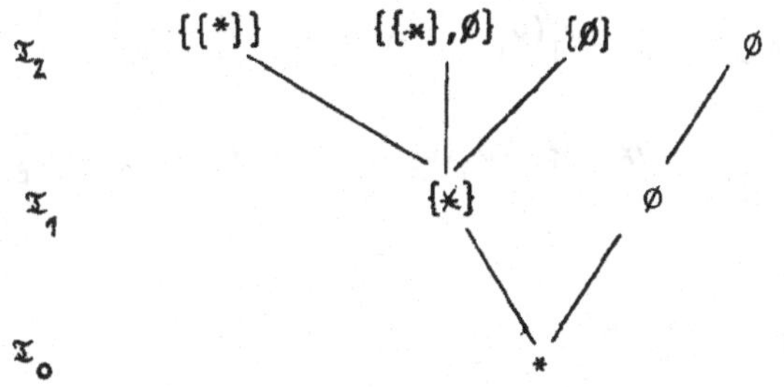

1.32 <u>Remarks.</u> a) The tree just considered has no maximal points: For $\alpha \in \Sigma_m$ and $m \leq n$ there is $\beta \in \Sigma_n$ with $\alpha = (\beta)_m$.

b) The Σ_n' s are not disjoint. But if $\alpha \in \Sigma_m \cap \Sigma_n$ and $m \leq n$, then we have $\alpha = (\alpha)_m$ (see c). Thus, "a is of type α" is unambiguously defined.

c) The whole picture is: Let $\alpha \in \Sigma_m$ and assume that

(i) α contains *. Then $\alpha \notin \Sigma_n$ for $m \neq n$.

(ii) α does not contain *. Then for all $n \geq m$ and $\beta \in \Sigma_n$

$$(\beta)_m = \alpha \quad \text{iff} \quad \beta = \alpha \ .$$

d) If U is an open subspace of A and $a \in U$, we have $t_n(A,a) = t_n(U,a)$.

We want to characterize up to L_t-equivalence a T_3-space by the types of its points.

1.33 <u>Definition.</u> Given $n \in \omega$ and any topological space (A,σ), we define the function $K_n^{(A,\sigma)}: \Sigma_n \to \omega \cup \{\infty\}$ by

$$K_n^{(A,\sigma)}(\alpha) = \text{number of } a \in A \text{ with } t_n(a) = \alpha \ .$$

If the topology σ is understood, then we shall not mention it explicitly.- By 1.32 b), $K^A = UK_n^A$ is a function on Σ.

1.34 <u>Theorem.</u> Two T_3-spaces (A,σ) and (B,τ) are L_t-equivalent iff $K^A = K^B$.

<u>Proof.</u> For one direction, we define formulas $\varphi_\alpha^n(v_o)$ such that for any space (A,σ) and $a \in A$,

$$(A,\sigma) \models \varphi_\alpha^n[a] \quad \text{iff} \quad t_n(A,a) = \alpha \ .$$

Put $\varphi_*^o(v_o) = v_o = v_o$ and for $\alpha \in \Sigma_{n+1}$,

$$\varphi_\alpha^{n+1}(v_o) = \bigwedge_{\beta \in \alpha} \forall X \ni v_o \ \exists v_1 (v_1 \in X \wedge \neg v_1 = v_o \wedge \varphi_\beta^n(v_1)) \wedge$$

$$\bigwedge_{\beta \in \Sigma_n \setminus \alpha} \neg \forall X \ni v_o \ \exists v_1 (v_1 \in X \wedge \neg v_1 = v_o \wedge \varphi_\beta^n(v_1)).$$

Whence

$$K_n^A(\alpha) \geq k \quad \text{iff} \quad (A,\sigma) \models \exists v_0 \ldots \exists v_{k-1} (\bigwedge_{i<k} \varphi_\alpha^n(v_0) \wedge \bigwedge_{i \neq j} \neg v_i = v_j).$$

Therefore, $(A,\sigma) \equiv^t (B,\tau)$ implies $K_n^A = K_n^B$.

For the proof of the other direction we need the following lemma.

1.35 <u>Lemma</u>. Let ξ be an ordinal and $(R_\eta)_{\eta<\xi}$ a family of symmetric relations between (possibly empty) topological spaces $(A,a_1,\ldots,a_m)$ with a finite set of distinguished points. Suppose that whenever

$$(A,a_1,\ldots,a_m) R_\eta (B,b_1,\ldots,b_m) \quad \text{and} \quad \eta' < \eta$$

conditions 1) and 2) are satisfied.

1) For all $a_{m+1} \in A \setminus \{a_1,\ldots,a_m\}$ there is $b_{m+1} \in B$ s.t.
$(A,a_1,\ldots,a_{m+1}) R_{\eta'} (B,b_1,\ldots,b_{m+1})$.

2) For every $i = 1,\ldots,m$ and every neighborhood U' of a_i there are clopen sets U and V s.t.
$a_i \in U \subset U' \setminus \{a_1,\ldots,\hat{a}_i,\ldots,a_m\}$ (a_i omitted), $b_i \in V \subset B \setminus \{b_1,\ldots,\hat{b}_i,\ldots,b_m\}$,
$(A \setminus U,a_1,\ldots,\hat{a}_i,\ldots,a_m) R_{\eta'} (B \setminus V,b_1,\ldots,\hat{b}_i,\ldots,b_m)$ and $(U,a_i) R_{\eta'} (V,b_i)$.

Then $(A,a_1,\ldots,a_m) \simeq_\xi^t (B,b_1,\ldots,b_m)$, if $(A,a_1,\ldots,a_m) R_\eta (B,b_1,\ldots,b_m)$ holds for all $\eta < \xi$. (For the definition of $\simeq_\xi^t$ see I.4.9).

<u>Proof</u>. For $\eta < \xi$ let I_η be the set of all triples

$$(\{(a_j^i,b_j^i) \mid i \leq k, j = 1,\ldots,m_i\}, \{(\bigcup_{i\in s} U^i, \bigcup_{i\in s} V^i) \mid s \subset \{0,\ldots,k\}\},$$
$$\{(\bigcup_{i\in s} U^i, \bigcup_{i\in s} V^i) \mid s \subset \{0,\ldots,k\}\}),$$

where $(U^i)_{i\leq k}$ is a clopen partition of A,

$(V^i)_{i\leq k}$ is a clopen partition of B and

$$(U^i,a_1^i,\ldots,a_{m_i}^i) R_\eta (V^i,b_1^i,\ldots,b_{m_i}^i) \quad \text{for } i \leq k.$$

It is easy to see that $(I_\eta)_{\eta<\xi}$ has the properties (forth_1), (back_1), (forth^*) and (back^*) (see p. 16, 19).

Our conclusion follows from the fact that

$(A,a_1,\dots,a_m)R_\eta(B,b_1,\dots,b_m)$ implies $(\{(a_j,b_j) \mid j = 1,\dots,m\},$
$\{(A,B),(\emptyset,\emptyset)\}, \{(A,B),(\emptyset,\emptyset)\}) \in I_\eta$.

We apply 1.35 for $\xi = \omega$ and define $(R_n)_{n<\omega}$ by

$(A,a_1,\dots,a_m)R_o(B,b_1,\dots,b_m)$ iff A,B are (possibly empty) T_3-spaces with clopen bases, and $a_i \neq a_j$, $b_i \neq b_j$ for $i \neq j$,

$(A,a_1,\dots,a_m)R_{n+1}(B,b_1,\dots,b_m)$ iff $(A,a_1,\dots,a_m)R_o(B,b_1,\dots,b_m)$,
$t_n(A,a_i) = t_n(B,b_i)$ for $i = 1,\dots,m$
and $K_n^A | n + m + 1 = K_n^B | n + m + 1$

(here $f|m$ denotes the function $\min(f(x),m)$).

Then $R_{n+1} \subset R_n$. Suppose $(A,a_1,\dots,a_m)R_{n+1}(B,b_1,\dots,b_m)$. We want to prove 1),2) of 1.35 for $\eta = n+1$, $\eta' = n$.

1) Let $a_{m+1} \in A \setminus \{a_1,\dots,a_m\}$. Let k be the number of indices $i \in \{1,\dots,m\}$ s.t. $t_n(a_i) = t_n(a_{m+1}) = \alpha$. Then $K_n^A(\alpha) > k$, and therefore $K_n^B(\alpha) > k$. Since $t_n(a_i) = \alpha$ iff $t_n(b_i) = \alpha$, there is $b_{m+1} \in B \setminus \{b_1,\dots,b_m\}$ with $t_n(b_{m+1}) = \alpha$. We have $(A,a_1,\dots,a_{m+1})R_n(B,b_1,\dots,b_{n+1})$, for t_{n-1} is determined by t_n, and $K_{n-1}^C | r$ is determined by $K_n^C | r$.

2) Let U' be a neighborhood of a_i. We always find a clopen neighborhood U of a_i, $U \subset U'$ s.t.

a) $a_j \notin U$ for $j \neq i$, and if $n > 0$

b) $c \in U \setminus \{a_i\}$ implies $t_{n-1}(c) \in t_n(a_i)$,

c) $K_{n-1}^A(\alpha) = \infty$ implies $K_{n-1}^{A-U}(\alpha) \geq n+m$.

If we choose the neighborhood V in B in the same way, we have $(A \setminus U,a_1,\dots,\hat{a}_i,\dots,a_m)R_n(B \setminus V,b_1,\dots,\hat{b}_i,\dots,b_m)$ and $(U,a_i)R_n(V,b_i)$. For we have for example,

$$t_{n-1}(U,a_i) = (t_n(A,a_i))_{n-1} = (t_n(B,b_i))_{n-1} = t_{n-1}(V,b_i)) ,$$

and $K_{n-1}^U | n + m$, $K_{n-1}^{A-U} | n + m$ resp. $K_{n-1}^V | n + m$, $K_{n-1}^{B-V} | n + m$ are completely determined by $t_n(a_i)$ and $K_{n-1}^A | n + m + 1$ resp. $t_n(b_i)$ and $K_{n-1}^B | n + m + 1$.

More precisely, a), b) and c) imply:

$$K_{n-1}^{U}(\alpha) = \begin{cases} \infty, & \text{if } \alpha \in t_n(a_i) \\ 1, & \text{if } \alpha = t_{n-1}(a_i) \notin t_n(a_i) \\ 0, & \text{otherwise .} \end{cases}$$

$$K_{n-1}^{A-U}(\alpha)|n+m = \begin{cases} K_{n-1}^{A}(\alpha)|n+m & , \quad \text{if } \alpha \neq t_{n-1}(a_i) \\ (K_{n-1}^{A}(\alpha)|n+m+1)-1, & \text{otherwise.} \end{cases}$$

Now we complete the proof of 1.34 as follows. Suppose $K^A = K^B$. By 1.17 there are T_3-spaces A' and B' with clopen bases s.t. $A' \equiv^t A$ and $B' \equiv^t B$. Since $K^{A'} = K^{B'}$, we have $A'R_nB'$ for all $n < \omega$. Thus, by 1.35, $A' \simeq^t_\omega B'$, and by I.4.13 $A' \equiv^t B'$.

1.36 <u>Remarks and exercises.</u> a) We have shown that $A \simeq^t_{2n+k+1} B$ implies $K_n^A|k = K_n^B|k$,and that $K_n^A|n+1 = K_n^B|n+1$ implies $A \simeq^t_{n+2} B$.

b) The formulas φ_α^n are (equivalent to) Δ-formulas. Compare I.5.13 and 1.32 d). Also the φ_α^n and the formulas "$K_n(\alpha) \geq k$" are equivalent to $L(I)$-formulas (defined on p. 48). For "$\beta \in t_{n+1}(x)$" may be expressed by $\neg Ixy(\varphi_\beta^n(y) \rightarrow x = y)$.

c) Show that in T_3-spaces every L_t-sentence φ is equivalent to a disjunction of sentences of the form "$K_n|n+1 = h$". It is possible to obtain this sentences in an effective way by retracing the proof of 1.34.

d) Show that in T_3-spaces every L_t-formula $\varphi(x_1,\ldots,x_n)$ is equivalent to a disjunction of formulas of the form "$K_n|n+1 = h$", "$t_n(x_i) = \alpha$", $x_i = x_j$, $\neg x_i = x_j$.

Compare 1.14 b). It is not true that in ω-separated spaces every L_t-formula $\varphi(x_1,\ldots,x_n)$ is equivalent to a boolean combination of sentences, Δ-formulas $\psi(x_i)$ and equalities $x_i = x_j$.

e) Show that in T_3-spaces every Δ-formula $\varphi(x)$ is equivalent to a formula of the form "$t_n(x) \in s$", where $s \subset \mathfrak{T}_n$.

f) Let $S(\mathfrak{B})$ denote the Stone space of the Boolean algebra $\mathfrak{B}$. Show that $S(\mathfrak{B}) \equiv^t S(\mathfrak{B}')$ implies $\mathfrak{B} \equiv_{L_{\omega\omega}} \mathfrak{B}'$, but the converse does not hold in general.

We want to determine, what n-types can occur in T_3-spaces. For example,

$\{\{\emptyset\}\} \in \mathfrak{X}_3$ can never be the 3-type of a point a, because, if a is accumulation point of points of type $\{\emptyset\}$, it must also be accumulation point of isolated points. (It is not clear if the non-realizablity of a type always has such a simple reason.)

First we give a general method to construct T_3-spaces. Let (A,σ) be a T_3-space. If $B \subset A$ and a is an accumulation point of B, we write $B \to a$. $B \to C$ means that $B \to c$ for all $c \notin C$. $B \not\to C$ means $B \not\to c$ for all $c \in C$.

1.37 Definition. A partition $(A_p)_{p \in P}$ of A is good, if

(1) $A_p \to A_q$ or $A_p \not\to A_q$ for all $p,q \in P$.

(2) $\bigcup\{A_p \mid p \in P, A_p \not\to A_q\} \not\to A_q$ for all $q \in P$.

For finite P, (2) follows from (1). $A_p \to A_q$ induces a transitive binary relation on P. Note that A_p is infinite, if there is a $q \in P$ with $A_p \to A_q$.

1.38 Lemma. Let $\prec$ be a transitive binary relation on P. Suppose that $K:P \to \{n \mid n \geq 1\} \cup \{\infty\}$ is a function with $K(p) = \infty$ for non-minimal elements p (i.e. there is $q \in P$ with $q \prec p$). Then there is a T_3-space (A,σ) and a good partition $(A_p)_{p \in P}$ of A s.t.

$$A_p \to A_q \quad \text{iff} \quad q \prec p \quad \text{and} \quad |A_p| = K(p)$$

for all $p,q \in P$. For denumerable P, we find a denumerable (A,σ). We call (A,σ) a $(P,\prec,K)$-space or, more briefly, a $(P,\prec)$-space.

Proof. We construct disjoint sets A_p^n for $p \in P$ and $n \in \omega$ by induction on n. Choose A_p^o s.t. $|A_p^o| = K(p)$. If for $n \in \omega$, all A_p^n, $p \in P$, are defined, choose infinite disjoint sets $A_{q,a}^n$ for $q \in P$, $a \in \bigcup_{p \in P} A_p^n$. Set

$$A_q^{n+1} = \dot{\bigcup}\{A_{q,a}^n \mid a \in A_p^n \text{ and } p \prec q \text{ for some } p\}.$$

$T = \dot{\bigcup}_{\substack{p \in P \\ n \in \omega}} A_p^n$ has an ω-tree structure $\leq$ (but is possibly uncountable) defined by:

for $a \in A_p^n$: $N(a)$ = immediate successors of $a = \bigcup_{p \prec q} A_{q,a}^n$.

We set $(A,\sigma) = (T,\tau_{\leq})$ and $A_p = \bigcup_{n \in \omega} A_p^n$.

The next lemma shows that the elementary type of (A,σ) is determined by $(P,\prec,K)$.

1.39 Definition. Let $\prec$ be a transitive relation on P. Define $s_n:P \to \mathfrak{X}_n$ by

$$s_o(p) = * , \quad s_{n+1}(p) = \{s_n(q) \mid p \prec q\}.$$

Sometimes we write $s_n(P,q)$ for $s_n(p)$. Note that the remarks 1.32 hold for s_n instead of t_n; e.g. if $m \leq n$ then $s_m(p) = (s_n(p))_m$. And $s_n(P,p) = s_n(\{q \mid q \in P, p \preceq q\}, p)$.

1.40 Lemma. Let (A,σ) be a $(P,\prec,K)$-space with the good partition $(A_p)_{p \in P}$. Then

a) $t_n(a) = s_n(p)$ for $p \in P$ and $a \in A_p$.

b) $K_n^A(\alpha) = \sum_{s_n(p) = \alpha} K(p)$ for $\alpha \in \mathfrak{T}_n$.

The proof by induction on n is easy.

Our problem, what types occur in T_3-spaces, is solved by:

1.41 Theorem. For any function $h: \mathfrak{T}_n \to \omega \cup \{\infty\}$ the following are equivalent:

(i) There is a T_3-space (A,σ) with $K_n^A = h$.

(ii) There is a transitive relation $\prec$ on

$$\mathbb{V} = \{\alpha \in \mathfrak{T}_n \mid h(\alpha) \neq 0\}$$

s.t. for all $\alpha \in \mathbb{V}$, $\alpha = \{(\beta)_{n-1} \mid \beta \in \mathbb{V}, \alpha \prec \beta\}$, and α non-minimal implies $h(\alpha) = \infty$.

Proof. Assume (ii). Set $(P,\prec) = (\mathbb{V},\prec)$ and $K = h|_{\mathbb{V}}$.
Choose – using 1.38 – a $(P,\prec,K)$-space (A,σ). We compute K_n^A: By the assumption on $\prec$, we have

$$(\alpha)_{m+1} = \{(\beta)_m \mid \alpha \prec \beta\} \quad \text{for } m < n.$$

Using this we obtain by induction

$$s_{m+1}(\alpha) = \{s_m(\beta) \mid \alpha \prec \beta\} = \{(\beta)_m \mid \alpha \prec \beta\} = (\alpha)_{m+1} .$$

Whence $s_n(\alpha) = \alpha$ and

$$K_n^A(\alpha) = \sum_{s_n(\beta) = \alpha} K(\beta) = h(\alpha).$$

On the other hand, let (A,σ) be a T_3-space, and $K_n^A = h$. We denote by A_α the set of points of type α. We define the binary relation $\vdash$ on $\mathbb{V}$ by

$$\alpha \vdash \beta \quad \text{iff} \quad A_\beta \to a \quad \text{for some } a \in A_\alpha .$$

Claim 1. $\alpha = \{(\beta)_{n-1} | \alpha \vdash \beta\}$.

Proof: Clearly, $\alpha \vdash \beta$ implies $(\beta)_{n-1} \in \alpha$. Conversely, if $\gamma \in \alpha = t_n(a)$, then we have $A_\gamma \to \alpha$. But
$A_\gamma = \bigcup\{A_\beta | (\beta)_{n-1} = \gamma,\ \beta \in \mathfrak{T}_n\}$, thus $A_\beta \to a$, $(\beta)_{n-1} = \gamma$ for some $\beta \in \mathfrak{T}_n$. Then $\alpha \vdash \beta$.

Claim 2. $\alpha \vdash \beta$ implies $\beta \subset \alpha$.

Proof: Assume $A_\beta \to a$, $a \in A_\alpha$ and $\gamma \in \beta$. Then $A_\gamma \to A_\beta$, therefore $A_\gamma \to a$, i.e. $\gamma \in \alpha$.

Denote by $\prec$ the transitive closure of $\vdash$ on $\mathbb{V}$.

Claim 3. $\{(\beta)_{n-1} | \alpha \vdash \beta\} = \{(\beta)_{n-1} | \alpha \prec \beta\}$.

Proof. The inclusion $\subset$ is clear, since $\alpha \vdash \beta$ implies $\alpha \prec \beta$. If $\alpha \prec \beta$, then there is a sequence $\alpha_1,\dots,\alpha_k$ s.t. $\alpha \vdash \alpha_1 \vdash \dots \vdash \alpha_k \vdash \beta$. Then $(\beta)_{n-1} \in \alpha_k \subset \dots \subset \alpha_1 \subset \alpha$.

By claims 1 and 3, we have $\alpha = \{(\beta)_{n-1} | \alpha \prec \beta\}$. If $\alpha \in \mathbb{V}$ is not minimal, there is a β with $\beta \vdash \alpha$. Thus A_α has an accumulation point in A_β and $h(\alpha) = |A_\alpha| = \infty$.

As a corollary we obtain for $\gamma \in \mathfrak{T}_n$ the equivalence of (i) - (iii):

(i) γ is the type of a point of some T_3-space .

(ii) There is a set $\mathbb{V}$ with $\gamma \in \mathbb{V} \subset \mathfrak{T}_n$, and a transitive relation $\prec$ on $\mathbb{V}$ s.t. $\alpha = \{(\beta)_{n-1} | \alpha \prec \beta\}$ for all $\alpha \in \mathbb{V}$.

(iii) There is a (finite) transitive relation $(P,\prec)$ s.t. $\gamma = s_n(p)$ for some $p \in P$.

Theorem 1.41 yields a decision procedure for the theory of T_3-space:

Let φ be given. Look for functions $h_1,\dots,h_k : \mathfrak{T}_n \to \{0,\dots,n+1\}$ s.t.

$$(A,\sigma) \models \varphi \quad \text{iff} \quad K_n^A | n+1 \in \{h_1,\dots,h_k\} \qquad \text{(cf. 1.36 c)}$$

Set $\mathbb{V}_i = \{\alpha | h_i(\alpha) \neq 0\}$ and $U_i = \{\alpha | h_i(\alpha) = n+1\}$.

Then φ is satisfiable iff for some i there is a transitive relation $\prec$ on $\mathbb{V}_i$ s.t. for all $\alpha \in \mathbb{V}_i$,

$\alpha = \{(\beta)_{n-1} | \alpha < \beta\}$ and $\alpha \in U_i$ for non-minimal α.

In a similar way, one obtains decision procedures for the theory of regular spaces or for the theory of T_3-spaces with unary relations.

1.42 Remarks and exercises.

a) 1.41 also holds (with the same proof) for T_1-spaces.

b) Show that very T_3-space (A,σ) is L_t-equivalent to a T_3-space with a good partition. There are two proofs:

1) Set $L' = \{\approx\}$. The proof of 1.41 shows that

$$Th_t((A,\sigma)) \cup \text{"}\approx \text{ is a good equivalence relation"}$$

is finitely satisfiable.

2) We can assume that (A,α) is recursively saturated and define $\approx$ by

$$a \approx b \quad \text{iff} \quad t_n(a) = t_n(b) \quad \text{for all } n \in \omega .$$

c) Determine the elementary types of all structures $(A,\nu_1,\ldots,\nu_k)$, where ν_i is a filter on A, and prove the decidability of the corresponding theory(cf. p. 56).

D Finitely axiomatizable and $\aleph_o$-categorical T_3-spaces.

1.43 Defintion. Let (A,σ) be a T_3-space. The ∞-type of a point $a \in A$ is the sequence $t(a) = (t_n(a))_{n \in \omega}$. (A,σ) is of finite type, if $\{t(a) | a \in A\}$ is finite.

Note that (A,σ) is of finite type iff for some $n \in \omega$, $t_n(a) = t_n(b)$ implies $t_m(a) = t_m(b)$ for all $m \in \omega$ and $a,b \in A$. Consequently, if (A,σ) is of finite type, every space L_t-equivalent to (A,σ) is of finite type too.

1.44 Lemma. (A,σ) is of finite type iff (A,σ) has a finite good partition.

Proof. If $(A_p)_{p \in P}$ is a good partition and $<$ on P is defined by

$$q < p \quad \text{iff} \quad A_p \to A_q,$$

then we have $t(a) = (s_n(p))_{n \in \omega} = s(p)$ for $a \in A_p$.
Thus $\{t(a) | a \in \omega\}$ is finite, if P is.

Conversely, let (A,σ) be of finite type. Choose $n \in \omega$ s.t. $t(a)$ only depends on $t_n(a)$. Set $P = \{t_n(a) | a \in A\}$, and $A_\alpha = \{a | t_n(a) = \alpha\}$. This a good partition, since $A_\beta \to a$ is equivalent to $\beta \in t_{n+1}(a)$.

Note that under the hypothesis and notations of the second part of the proof, if $(B,\tau) \equiv^t (A,\sigma)$ the sets $B_\alpha = \{b | t_n(b) = \alpha\}$ for $\alpha \in P$ yield a good partition of (B,τ). We have

$$B_\beta \to B_\alpha \quad \text{iff} \quad A_\beta \to A_\alpha \quad \text{and} \quad |B_\alpha| = |A_\alpha|.$$

The following exercise shows that in the above proof it is enough to choose $n \geq 2|\{t(a) | a \in A\}| - 1$.

<u>Exercise</u>. Show that for all transitive relations $(P,\lessdot)$ and $(Q,\lessdot)$ and for $n \geq |P| + |Q| - 1$,

$$s_n(P,p) = s_n(Q,q) \quad \text{implies} \quad s_{n+1}(P,p) = s_{n+1}(Q,q)$$

(use induction on $|P| + |Q|$).

We call a topological structure $(\mathfrak{A},\sigma)$ <u>finitely axiomatizable</u>, if $Th_t((\mathfrak{A},\sigma))$ is finitely axiomatizable.

1.45 <u>Theorem</u>. The finitely axiomatizable T_3-spaces are just the $(P,\lessdot,K)$-spaces, where P is finite and where $K(p)$ is finite for any minimal element $p \in P$.

<u>Proof</u>. Let (A,σ) be a $(P,<,K)$-space with finite P and $K(p)$ finite for minimal $p \in P$. Then (A,σ) is of finite type. Choose n large enough s.t. $t_n(a)$ determines $t(a)$ for $a \in A$. We introduce the notations $m_\alpha = K^A_{n+1}(\alpha)$, $W = \{\alpha \in \mathfrak{T}_{n+1} | m_\alpha < \infty\}$ and $U = \{\alpha \in \mathfrak{T}_{n+1} | m_\alpha = \infty\}$. (A,α) is a model of the finite theory

$$T = \text{"}T_3 + \bigwedge_{\alpha \in W} K^{\cdots}_{n+1}(\alpha) = m_\alpha + \bigwedge_{\alpha \in U} K^{\cdots}_{n+1}(\alpha) \neq 0\text{"} .$$

We want to show that all models of T are L_t-equivalent. By 1.40 and 1.34 it is enough to show that all models (B,τ) of T are $(\mathbb{V},\lessdot^*,K^*)$-spaces, where $\mathbb{V} = \{\alpha \in \mathfrak{T}_{n+1} | m_\alpha \neq 0\}$, $K^*(\alpha) = m_\alpha$ and $\alpha \lessdot^* \beta$ iff $(\beta)_n \in \alpha$.
We have

$$K^B_{n+1}(\alpha) \neq 0 \quad \text{iff} \quad \alpha \in \mathbb{V}.$$

Thus $(B_\alpha)_{\alpha \in \mathbb{V}}$ is a partition of B, where $B_\alpha = \{b | t_{n+1}(b) = \alpha\}$.

Let $b \in B_\alpha$. Then, $B_\beta \to b$ iff $(\beta)_n \in \alpha$. For, if $B_\beta \to b$, then, clearly, $(\beta)_n \in \alpha$. And if $(\beta)_n \in \alpha$, then there is a $\gamma \in \mathbb{V}$ with $(\gamma)_n = (\beta)_n$ and $B_\gamma \to b$ (cf. proof of 1.41, claim 1). But any $\delta \in \mathbb{V}$ is determined by $(\delta)_n$. Thus $\gamma = \beta$ and $B_\beta \to b$.
It remains to show that $|B_\alpha| = m_\alpha$. This follows immediately from $(B,\tau) \models T$, if $m_\alpha < \infty$. If $m_\alpha = \infty$, there is a $p \in P$ with $s_{n+1}(p) = \alpha$ and $K(p) = \infty$. p is not minimal, so there is a $q \in P$ with $q \prec p$. But now $(\alpha)_n \in s_{n+1}(q) \in \mathbb{V}$, and therefore $B_\alpha \to B_{s_{n+1}(q)}$. Thus B_α must be infinite. Note that $K^*(\alpha)$ is finite, if α is $\prec^*$-minimal.

Conversely, suppose that (A,σ) is finitely axiomatizable. Using 1.34 we see that there is a $n \in \omega$ s.t.

$$(B,\tau) \equiv^t (A,\sigma) \quad \text{iff} \quad K_n^B| n + 1 = K_n^A|n + 1.$$

By 1.41 there is a transitive relation $\prec$ on

$\mathbb{V} = \{\alpha | K_n^A(\alpha) \neq 0\}$ s.t. $\alpha = \{(\beta)_{n-1} | \alpha \prec \beta, \beta \in \mathbb{V}\}$ and $K_n^A(\alpha) = \infty$ for non-minimal α.

1.38 yields a $(\mathbb{V},\prec,h)$-space (B,τ), where

$$h(\alpha) = \begin{cases} K_n^A| n + 1(\alpha), & \text{if } \alpha \text{ is minimal} \\ \infty, & \text{otherwise} \end{cases}$$

The proof of 1.41 shows that $K_n^B = h$. Whence $(B,\tau) \equiv^t (A,\sigma)$. Now we apply the first part of our proof to (B,τ) and conclude that (A,σ) is of the required form.

As a corollary we obtain from the preceding proof that every sentence, which is satisfiable in a T_3-space, also is satisfied in a finitely axiomatizable space.

1.46 <u>Exercise</u>. Let $\prec$ be transitive on P and $n \geq 2|P|$. Then for any point a of a T_3-space (A,σ),

$$t_n(a) = s_n(p) \quad \text{implies} \quad t_{n+1}(a) = s_{n+1}(p).$$

Whether a $(P,\prec,K)$-space and a $(P^*,\prec^*,K^*)$-space are L_t-equivalent can be seen (using 1.40 and 1.45) by a computation, which may be rather long. To make it easier we introduce the following concept:

1.47 <u>Definition</u>. Let $<$ be a transitive relation on P. (P, $<$) is <u>normal</u>, if for all $p,q \in P$, $p \neq q$, one of the following four properties (1) - (4) hold.

(1) $q \nless q$, $q \nless p$, $p < p$

(2) $p \nless p$, $p \nless q$, $q < q$

(3) $p < r$, $q \nless r$ for some $r \neq p$

(4) $q < r$, $p \nless r$ for some $r \neq q$.

The following picture gives an example of a normal (P,$<$). For distinct p and q, we have $p < q$ iff we can reach q from p on ascending lines. A point p is marked by ○ iff $p < p$.

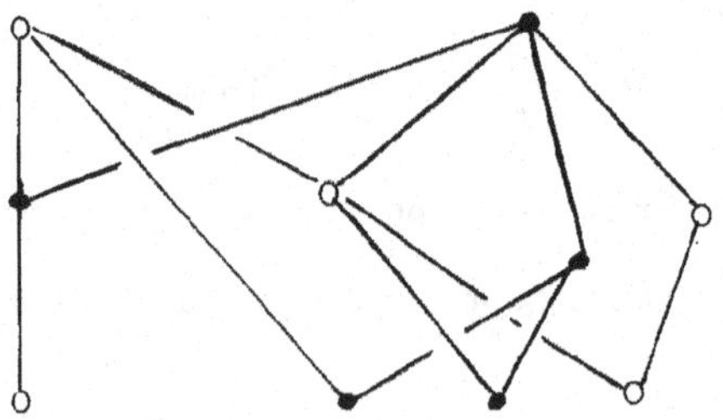

Note that in a normal P every final segment $Q \subset P$ (i.e. $q \in Q$ and $q < p$ imply $p \in Q$) is normal too.

1.48 <u>Lemma</u>. Let P be finite and $<$ transitive on P.
Then

(P,$<$) is normal iff $s(p) = s(q)$ implies $p = q$ for all $p, q \in P$.

(as in 1.43, $s(p)$ denotes $(s_n(p))_{n \in \omega}$).

<u>Proof</u>. Let (P,$<$) be normal. We show

(*) if $s(p) = s(q)$ then $p = q$

by induction on $|p| + |q|$, where $|p| = |\{r \mid p < r, p \neq r\}|$. Assume $s(p) = s(q)$. We show that none of (1) - (4) of 1.47 hold, thus obtaining $p = q$. To show that (1) of 1.47 fails assume that $q \nless q$ and $p < p$. Then $s_n(p) \in s_{n+1}(p) = s_{n+1}(q)$ for all n. Since P is finite, there is a t with $q < t$ and $s(t) = s(p)$. We have $|t| < |q|$ and hence, $p = t$ by induction. Thus (1) is false. By the same reasons, (2) does not hold. For (3), assume $p < r$ for some $r \neq p$. As above, there is t with $q < t$ and $s(t) = s(r)$. By induction hypothesis, $t = r$ and therefore (3) fails. Similarly one shows that (4) does not hold.

Conversely, assume that (*) holds. Let $p,q \in P$, $p \neq q$. Choose the smallest m s.t. $s_{m+1}(p) \neq s_{m+1}(q)$. W.l.o.g. suppose $s_{m+1}(p) \not\subset s_{m+1}(q)$. Then there is an r with $p \prec r$ and $s_m(r) \notin s_{m+1}(q)$. We have $q \not\prec r$. Thus (3) holds or $r = p$. In the latter case, we have $s_m(r) = s_m(q)$ and therefore, $q \not\prec q$. This shows that (1) holds.

1.49 Theorem. Every space (A,σ) of finite type is a $(P,\prec,K)$-space for a finite normal $(P,\prec)$. $(P,\prec,K)$ is uniquely determined by $Th_t((A,\sigma))$.

Proof. Let (A,σ) be of finite type and let $(A_\alpha)_{\alpha \in \mathbb{V}}$ be the good partition we used in the proof of 1.44 .
Define $\prec$ on $\mathbb{V}$ by

$$\alpha \prec \beta \quad \text{iff} \quad A_\beta \to A_\alpha .$$

Then $(\mathbb{V},\prec)$ is normal, since for $a \in A_\alpha$

$s_n(\mathbb{V},\alpha) = t_n(a) = \alpha$.

If (A,σ) is a $(P^*,\prec^*,K^*)$-space, we can find $(\mathbb{V},\prec,K)$ by

$$\mathbb{V} = \{s_n(p) \mid p \in P^*\}$$

$$s_n(p) \prec \alpha \quad \text{iff} \quad p \prec^* q \quad \text{for some } q \in P^* \text{ with } s_n(q) = \alpha$$

$$K(\alpha) = \sum_{s_n(p) = \alpha} K^*(p) .$$

If $(P^*,\prec^*)$ is normal and n is large enough

$$s_n(P^*,-):P^* \to \mathbb{V}$$

yields an isomorphism of $(P^*,\prec^*,K^*)$ onto $(\mathbb{V},\prec,K)$. But $(\mathbb{V},\prec,K)$ is uniquely determined by $Th_t((A,\sigma))$. - Note that if $(P,\prec)$ is normal, there is just one partition $(A_p)_{p \in P}$, which turns (A,σ) into a $(P,\prec)$-space:
$A_p = \{a \mid t(a) = s(p)\}$.

1.50 Exercises. a) Let $(P,\prec)$ and $(Q,\prec)$ be normal with smallest element p resp. q. Then $s_n(p) = s_n(q)$ for some $n \geq 2|P|$ implies $(P,\prec) \cong (Q,\prec)$.
b) $P_n = \{0,1,\dots,n\}$ with $\prec_n = \{(0,0)\} \cup \{(i,j) \mid 0 \leq i < j \leq n\}$ is normal. Put $\alpha_n = s_{2n}(P_n,0)$. Show that for every set $F \subset \omega$ there is a T_3-space A with

$$K^A_{2n}(\alpha_n) \neq 0 \quad \text{iff} \quad n \in F.$$

Thus there are $2^{\aleph_0}$ complete theories of T_3-spaces.

c) There are $2^{\aleph_0}$ distinct ∞-types $t(a)$ of elements of T_3-spaces.

d) Let $(P,<)$ be finite and normal. A $(P,\leqslant,K)$-space is finitely axiomatizable iff $K(p)$ is finite for minimal $p \in P$.

e) Every finitely axiomatizable T_3-space is L_t-equivalent to the Stone space of a Boolean algebra.

Hint: Let (A,σ) be a $(P.<)$-space. Call a clopen $U \subset A$ p-small iff $(A_q \cap U \neq \emptyset \Leftrightarrow p \leq q)$. If we construct A from $(P,<,K)$ as in 1.38 and use the basis given in 1.20 we get a basis α of σ consisting of small sets s.t. the difference of any two elements of α is a finite disjoint union of elements of α. Let $\mathfrak{B}$ be the Boolean algebra generated by α. Then $S(\mathfrak{B}) \equiv^t (A,\sigma)$, if $(P,\leqslant,K)$ is as in 1.45, and $p < q < p$ implies $p = q$.

1.51 <u>Definition</u>. An infinite topological structure $(\mathfrak{A},\sigma)$ is <u>$\aleph_0$-categorical</u>, if all denumerable topological structures $(\mathfrak{B},\tau)$, which are L_t-equivalent to $(\mathfrak{A},\sigma)$, are homeomorphic.

Examples of $\aleph_0$-categorical structures: (a) $((\mathbb{R},<),\tau)$, the structure of the reals with its ordering and its topology. (b) $(\mathbb{Q},\sigma)$, the rationals with its topology (cf. 1.22). (c) Infinite discrete spaces. (d) Infinite T_3-spaces with points only of type $\emptyset$ or $\{\{*\},\emptyset\}$. (e) Structures of the form (A,σ,τ), where σ and τ are T_3-topologies on the infinite set A, which are relatively prime, i.e.

$$\emptyset \neq U \in \sigma \text{ and } \emptyset \neq V \in \tau \quad \text{imply} \quad \emptyset \neq U \cap V.$$

(f) Structures of the form $((A,B),\sigma)$, where σ is a T_3-topology without isolated points and $B \subset A$ is perfect and nowhere dense (cf. p. 73).

1.52 <u>Definition</u>. A transitive relation $<$ on P is <u>dense</u>, if for all $p,q \in P$ with $p < q$ there is an $r \in P$ s.t. $p < r < q$.

Examples of normal and dense $(P,<)$ are

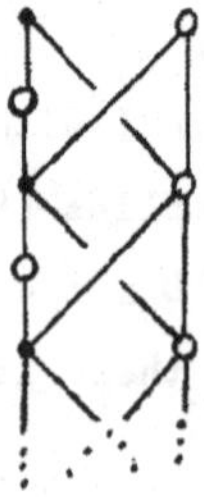

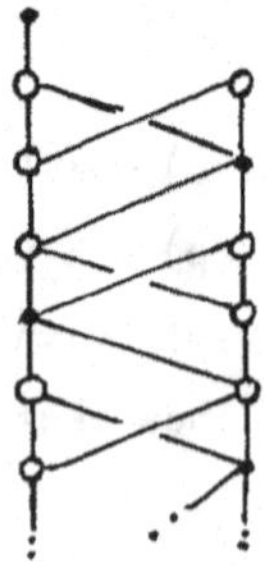

and the final segments 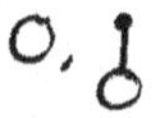 and 

(where points with $p < p$ are marked by $\circ$).

1.53 <u>Theorem</u>. The $\aleph_0$-categorical T_3-spaces are just the $(P,<)$-spaces, where $(P,<)$ is finite and dense. - Let (A,σ) be a $(P,<)$-space with finite and normal $(P,<)$. Then (A,σ) is $\aleph_0$-categorical iff $(P,<)$ is dense.

We divide the proof in four parts.

I. Let $(P^*,<^*)$ be finite and dense and $K^*:P^* \to \omega \cup \{\infty\}$. Then all denumerable $(P^*,<^*,K^*)$-spaces are homeomorphic.

Proof: We define a relation R between (possibly empty) topological spaces $(A,a_1,\ldots,a_m)$ with a finite number of distinguished points:

$$(A,a_1,\ldots,a_m)R(B,b_1,\ldots,b_m) \quad \text{iff}$$

a) $a_i \neq a_j$ and $b_i \neq b_j$ for $i \neq j$. A,B are (possibly empty) T_3-spaces with clopen bases.

b) For some finite and dense $(P,<)$ and $K:P \to \omega \cup \{\infty\}$, A and B are $(P,<,K)$-spaces with good partitions $(A_p)_{p \in P}$ resp. $(B_p)_{p \in P}$ with the property

$$a_i \in A_p \quad \text{iff} \quad b_i \in B_p.$$

(We allow $P = \emptyset$).

We show that R satisfies analogues of the back and forth properties of 1.35. Suppose $(A,a_1,\ldots,a_m)R(B,b_1,\ldots,b_m)$ and let $(P,<,K),(A_p)_{p \in P}$ and $(B_p)_{p \in P}$ be as above.

1) For all $a_{m+1} \in A\setminus\{a_1,\ldots,a_m\}$ there is a $b_{m+1} \in B$ s.t.

$$(A,a_1,\ldots,a_{m+1})R(B,b_1,\ldots,b_{m+1}).$$

This holds simply because $|A_p| = |B_p| = K(p)$.

2) For all i and every neighborhood U' of a_i there are clopen sets U and V s.t.

$$a_i \in U \subset U'\setminus\{a_1,\ldots,\hat{a}_i,\ldots,a_m\},\ b_i \in V \subset B\setminus\{b_1,\ldots,\hat{b}_i,\ldots,b_m\},$$

$$(A\setminus U,a_1,\ldots,\hat{a}_i,\ldots,a_m)R(B\setminus V,b_1,\ldots,\hat{b}_i,\ldots,b_m) \quad \text{and} \quad (U,a_i)R(V,b_i)\ .$$

To prove 2) assume that $a_i \in A_q$, and choose U and V small enough s.t.

$q < p$ implies $A_p \setminus U \neq \emptyset$ and $B_p \setminus V \neq \emptyset$,

$q \not< p$ implies $A_p \cap (U \setminus \{a_i\}) = \emptyset$ and $B_p \cap (V \setminus \{b_i\}) = \emptyset$.

Set $Q_o = \{p | q < p\} \cup \{q\}$. $(A_p \cap U)_{p \in Q_o}$ resp. $(B_p \cap V)_{p \in Q_o}$ are good partitions of U resp. V. Define K_o by

$$K_o(p) = \begin{cases} \infty, & \text{if } q < p \\ 1, & \text{if } p = q \text{ and } q \not< p. \end{cases}$$

Then U and V are $(Q_o, <, K_o)$-spaces, and $(U, a_i)R(V, b_i)$ holds. Now set

$$Q_1 = \begin{cases} P & \text{, if } K(q) > 1 \\ P \setminus \{q\} & \text{, otherwise} \end{cases} .$$

$(A_p \setminus U)_{p \in Q_1}$ and $(B_p \setminus V)_{p \in Q_1}$ are good partitions of $(A \setminus U)$ resp. $(B \setminus V)$.

Suppose $q < p$. There is an element r s.t. $q < r < p$. Since $(A_p \setminus U) \to (A_r \setminus U)$ and $A_r \setminus U \neq \emptyset$, $A_p \setminus U$ is infinite. Similarly, $B_p \setminus U$ is infinite. This shows that $A \setminus U$ and $B \setminus V$ are $(Q_1, <, K_1)$-spaces, where

$$K_1(p) = \begin{cases} K(p) & , \text{ if } p \neq q \\ K(p) - 1, & \text{if } p = q \text{ and } q \in Q_1. \end{cases}$$

Now, if A and B are denumerable $(P^*, <^*, K^*)$-spaces, then ARB (cf. 1.17). 1.35 (or more precisely the proof of 1.35) shows that $A \cong^t_p B$. Thus $A \cong B$ (by I 4.3 (b)).

II. Let $(P^*, <^*)$ be finite and dense. Then every $(P^*, <^*)$-space is $\aleph_0$-categorical.

Proof: (cf. the proof of 1.49). Let B and C be denumerable spaces L_t-equivalent to the $(P^*, <^*, K^*)$-space A. Then B and C are $(V, <, K)$-spaces, where (for large enough n)

$$V = \{s_n(p) | p \in P^*\} ,$$

$$s_n(p) < \alpha \quad \text{iff} \quad p <^* q \quad \text{for some } q \in P^* \text{ with } s_n(q) = \alpha,$$

$$K(\alpha) = \sum_{s_n(p) = \alpha} K^*(p) .$$

It remains to show that $(V, <)$ is dense. Then, by I, $B \cong C$. But if

$s_n(p) < s_n(q)$, $p <^* q$, then there is $r \in P$ with $p <^* r <^* q$. Then $s_n(p) < s_n(r) < s_n(q)$.

III. $\aleph_0$-categorical T_3-spaces are of finite type.

Proof: If we take any standard proof of the Ryll-Nardzewski theorem (e.g. [20], p. 100), and replace the omitting types theorem by I 9.2, we obtain:

Let T be a denumerable L(I)-theory. If there are infinitely many modulo T non-equivalent L(I)-formulas $\varphi(x)$, then there is a type $\Phi(x)$ of L(I)-formulas, which is realized in some model of T and omitted in another one.

Now take as T the L(I)-theory of a T_3-space (A,σ), which is not of finite type. There are infinitely many modulo T non-equivalent L(I)-formulas of the form "$t_n(x) = \alpha$" (cf. 1.36 b). Therefore we obtain a type $\Phi(x)$ of L(I)-formulas, which is realized in (B,τ) and omitted in (C,ϱ), where both spaces are denumerable models of T.

To finish the proof of III we need the following lemma (which by the way shows that "T_3" is not expressible by an L(I)-sentence).

1.54 <u>Lemma</u>. Let $(\mathfrak{A},\sigma)$ be a denumerable T_1-structure. Then there is a T_3-topology $\bar{\sigma} \supset \sigma$ with a denumerable basis s.t. the elements of A satisfy in $(\mathfrak{A},\sigma)$ and in $(A,\bar{\sigma})$ the same L(I)-formulas.

<u>Proof</u>: Let $(A_i)_{i \in \omega}$ be a list of all L(I)-definable subsets of A. Take a T_3-topology $\bar{\sigma} \supset \sigma$ with denumerable basis s.t. for all $i \in \omega$,

$$\sigma\text{-interior of } A_i = \bar{\sigma}\text{-interior of } A_i \quad (\text{cf. } [5]).$$

Now the assertion of the lemma is easily obtained by induction on L(I)-formulas.

Proof of III (continued). τ and ϱ are T_1, for "T_1" can be expressed by $\forall x\, Ixy\, \neg\, x = y$. If we choose $\bar{\tau} \supset \tau$ and $\bar{\varrho} \supset \varrho$ according to the lemma, then $(B,\bar{\tau})$ and $(C,\bar{\varrho})$ still are L(I)-equivalent to (A,σ). $(B,\bar{\tau}) \not\cong (C,\bar{\varrho})$, since Φ is realized in $(B,\bar{\tau})$ and omitted in $(C,\bar{\varrho})$. But (cf. 1.36 b) $K^{(A,\sigma)} = K^{(B,\bar{\tau})} = K^{(C,\bar{\varrho})}$ and thus, $(A,\sigma) \equiv^t (B,\bar{\tau}) \equiv^t (C,\bar{\varrho})$. Therefore (A,σ) is not $\aleph_0$-categorical.

IV. Let $(P,<)$ be finite, normal, and not dense. Then no $(P,<)$-space is $\aleph_0$-categorical.

(Note that I - IV together with 1.49 prove the theorem.)

Proof: Let (A,σ) be the $(P,\lessdot,K)$-space constructed in the proof of 1.38. Let $T,\leq$ and A_p (for $p \in P$) be as in that proof. Choose $p,q \in P$, $p \lessdot q$, s.t. there is no r with $p \lessdot r$ and $r \lessdot q$. Then, for any $a \in A_p$,

$$a \leq b \text{ and } b \in A_q \text{ implies } b \in N(a).$$

Thus $(U_\emptyset(a) \setminus U_\Delta(a)) \cap A_q$ is finite for every finite Δ (cf. 1.20).
Since P is finite and normal, we have, for sufficiently large n, $A_r = \{a \mid t_n(a) = s_n(r)\}$. Therefore (A,σ) satisfies the following $(L_{\omega_1\omega})_t$- sentence

$$\varphi = \exists x(t_n(x) = s_n(p) \wedge \exists X \ni x\ \forall Y \ni x\ \text{"}\{y \mid y \in X \setminus Y \wedge t_n(y) = s_n(q)\} \text{ is finite"}) .$$

Let (C,γ) be an ω_1-saturated weak structure with $(C,\gamma) \equiv_{L_2} (A,\sigma)$. $(C,\tilde{\gamma})$ does not satisfy φ. For, if $t_n(c) = s_n(p)$ then c is an accumulation point of points of type $s_n(q)$. Let $U \in \gamma$ be a neighborhood of c. Then there are arbitrarily large finite sets of the form $\{d \mid d \in U \setminus V, t_n(d) = s_n(q)\}$, where $c \in V \in \gamma$. Hence this set is infinite for some $V \in \gamma$ with $c \in V$.

By the Löwenheim-Skolem theorem there is a denumerable (D,ϱ) with $(D,\varrho) \equiv^t (C,\tilde{\gamma})$ and $(D,\varrho) \not\models \varphi$. Then $(D,\varrho) \equiv^t (A,\sigma)$ but $(D,\varrho) \not\cong^t (A,\sigma)$.

1.55 <u>Exercises</u>. a) Given an infinite T_3-space (A,σ) which is not $\aleph_0$-categorical, there are denumerable spaces (B,τ) and (C,ϱ) L_t-equivalent to (A,σ) and a $b \in B$ s.t. for no $c \in C, t(B,b) = t(C,c)$.

b) Characterize the $\aleph_0$-categorical structures of the form $((A,B_1,\dots,B_n),\sigma)$, where $B_i \subset A$ and σ is a T_3-topology.

c) We call (A,σ) <u>locally-finite</u>, if every point has an open neighborhood of finite type. $(P,\lessdot)$ is locally finite, if $\{q \mid p \lessdot q\}$ is finite for every p.

Prove:

1) The analogues of 1.44 and 1.49 for local finiteness.

2) If $(P,\lessdot)$ is locally finite and dense and $K:P \to \omega \cup \{\infty\}$, then all denumer-

able $(P,\leq,K)$-spaces are homeomorphic. This density is a necessary condition, in case that $(P,\leq)$ is normal. We call such denumerable spaces <u>dense spaces</u>.

3) There is a unique dense space containing all dense spaces as open subspaces.

d) Give a complete axiomatization of $(\mathbb{R},\nu)$, where ν is the natural uniformity on the reals. Show that $(\mathbb{R},\nu) \equiv_{L^2_m} (\mathbb{Q},\nu|\mathbb{Q})$.

§ 2 Topological abelian groups.

The class of all topological abelian groups is axiomatizable in L_t, $L = \{0,+,-\}$ (cf. I § 2). It was shown in [7,23] that the $L_{\omega\omega}$-theory of abelian groups is decidable. Moreover, all elementary types (i.e. all complete $L_{\omega\omega}$-theories of abelian groups) have been characterized. We want to study the corresponding problems for topological abelian groups.

In this section all groups are supposed to be abelian. We denote groups by A,B,... , thus, in general, not mentioning the group operations.

Let B be a subgroup of the group A. $\{a + B | a \in A\}$ is the basis of a group topology τ on A. The τ-closure of $\{0\}$ is B. Therefore we can interpret the theory of all pairs (A,B) consisting of an abelian group A and a distinguished subgroup B over the theory of topological groups. The first theory is hereditarily undecidable (see [2]), hence so is the theory of topological groups.

The main results of this section are (see 2.3 for definitions):

(2.8) The theory of locally pure and torsionfree topological groups is decidable.

(2.11) The theory of algebraically complete topological groups is decidable.

(2.14) The theory of hausdorff topological groups is hereditarily undecidable.

2.1 Remark. [6] contains a complete analysis of the (L_t-)elementary types of locally pure, torsionfree topological groups. Moreover, the following two undecidability results are proved: The theory of hausdorff and locally pure topological groups and the theory of hausdorff, torsionfree and divisible groups are hereditarily undecidable (both results imply 2.14). Also compare the section about topological groups in [I.9].

The topology τ of a topological group (A,τ) ($=((A,0,+,-),\tau)$) is determined by the neighborhood filter ν of 0:
$a+\nu$ is the neighborhood filter of a. Therefore we identify (A,τ) and (A,ν) and use L_m for (A,ν) instead of L_t for (A,τ) (cf. I.8.8b 2). - Note that a monotone system ν on a group A turns A into a topological group iff the following L_m-sentences hold in (A,ν):

$$\forall X\ \forall Y\ \exists Z\quad Z \subset X \cap Y$$

$$\forall X\ \ 0 \in X$$

$$\forall X\ \exists Y\ \ Y-Y \subset X$$

If ν is a filter with a basis consisting of subgroups of A, then (A,ν) satisfies these axioms. In fact, a monotone structure (B,μ) is a topological group iff it is L_m-equivalent to such an (A,ν). For we have

2.2 Lemma. An ω_1-saturated topological group has a basis consisting of subgroups.

Proof. Suppose that (A,ν) is an ω_1-saturated topological group, i.e. that (A,α) is an ω_1-saturated two-sorted structure for some basis α of ν. ν is closed under countable intersections (cf. 1.18). Now, for every $U \in \nu$, we obtain - using the axioms of a topological group - a sequence $U_0 \supset U_1 \supset \ldots$ s.t. $U_i \in \alpha$ and $U_{i+1} - U_{i+1} \subset U_i \subset U$. Then $\bigcap_{i \in \omega} U_i$ is a subgroup of A and $\bigcap_{i\in\omega} U_i \in \nu$.

2.3 Definition. A topological group (A,ν) is locally pure, if for every $n \geq 0$ the following L_m-sentence holds in (A,ν):

$$\forall X\ \ \exists Y\ \forall x\ \ \exists y(n\cdot x \in Y \rightarrow (y \in X \wedge ny = nx)).$$

An algebraically complete group is a locally pure group, which satisfies for each $n > 0$,

$$\forall x(\forall X\ \ \exists y\ ny - x \in X \ \rightarrow \exists y\ ny = x).$$

For torsionfree groups, local pureness is nothing else than continuity of (partial) division by n (for $n \geq 0$). Algebraic completeness means besides being locally pure that nA is closed for every $n > 0$. A locally pure group with a countable basis is algebraically complete iff it is pure in its completion (we will not use this fact). (B is a pure subgroup of A iff $nB = B \cap nA$ for all $n \geq 1$).

If ν is a filter with a basis consisting of pure subgroups of A, then (A,ν) is locally pure.

2.4 Lemma. An ω_1-saturated locally pure group has a basis consisting of pure subgroups.

Proof. Let (A,ν) be an ω_1-saturated locally pure group. Choose a basis α of ν s.t. (A,α) is an ω_1-saturated two-sorted structure. For every $U \in \nu$ there is a sequence $(U_i)_{i \in \omega}$ s.t. $U_i \in \alpha$, $U_{i+1} - U_{i+1} \subset U_i \subset U$ and $nA \cap U_{i+1} \subset nU_i$ for $n \leq i$. $\bigcap_{i \in \omega} U_i$ is a pure subgroup of A with $\bigcap_{i \in \omega} U_i \in \nu$. For, if $n\,a \in \bigcap_{i \in \omega} U_i$, then the type $\Phi(x) = \{nx = nc \wedge x \in C_i \mid i \in \omega\}$ is finitely satisfiable in $((A,(U_i)_{i \in \omega},a)\alpha)$. Choose b satisfying Φ. Then $na = nb$ and $b \in \bigcap_{i \in \omega} U_i$.

2.5 Examples. a) A group with the indiscrete (= trivial) resp. discrete topology is locally pure resp. algebraically complete.

b) $(\mathbb{Z},\nu_{\mathbb{Z}})$, where $\nu_{\mathbb{Z}}$ has the basis $\{n\,\mathbb{Z} \mid n > 0\}$ is locally pure (but the $n\mathbb{Z}$ are not pure in $\mathbb{Z}$).

c) The direct sum of locally pure resp. algebraically complete groups (with the topology induced by the product topology) is locally pure resp. alg. complete.

2.6 Lemma. An ω_1-saturated locally pure group has a unique decomposition $(D,\lambda) \oplus (B,\mu)$, where λ is the indiscrete topology on D, and where (B,μ) is hausdorff, locally pure and ω_1-saturated.

Proof. Given an ω_1-saturated locally pure group (A,ν), $D = \overline{\{0\}}$ is a pure subgroup (since (A,ν) is locally pure) and is ω_1-saturated, since it is definable. Whence D is a direct factor (cf.[7]). Now set $(B,\mu) = (A,\nu)/_D$.

2.7 Lemma. (A,ν) is a hausdorff, locally pure and torsionfree group iff there is a linearly ordered group $(B,<)$ s.t. $(A,\nu) \equiv_{L_m} (B,\nu_<)$, where $\nu_<$

is the order-topology.

(Note that $(\mathbb{Z},\nu_{\mathbb{Z}})$ is hausdorff, locally pure and torsionfree, but $\nu_{\mathbb{Z}}$ is not induced by an ordering).

Proof. A linearly ordered group is locally pure (since $nx \in (-nb,nb)$ implies $x \in (-b,b)$).

Now assume that (A,ν) is hausdorff, locally pure and torsionfree. Choose an ω_1-saturated topological group (C,μ) with $(C,\mu) \equiv_{L_m} (A,\nu)$. μ has a basis γ of pure subgroups. Let (B,β) be a weak denumerable structure with $(B,\beta) \equiv_{L_2} (C,\gamma)$. Then $(B,\hat{\beta}) \equiv_{L_m} (A,\nu)$. $\hat{\beta}$ has a descending basis $B = U_o \supset U_1 \supset \ldots$ of pure subgroups. $B/_{U_i}$ is torsionfree and therefore $(B/_{U_i}, <_i)$ is a linearly ordered group for some linear order $<_i$. Since $\bigcap_{i \in \omega} U_i = \{0\}$, for $b \in B, b \neq 0, I(b)$ is well-defined by the requirement that

$$b \in U_{I(b)} \setminus U_{I(b)+1} .$$

An easy calculation shows that

$$b < c \quad \text{iff} \quad b \neq c,\ b + U_{I(b-c)} <_{I(b-c)} c + U_{I(b-c)}$$

defines a linear ordering $<$ on B. $(B,<)$ is a linearly ordered group. We have $\nu_< = \hat{\beta}$, since $U_{I(b)} \subset (-b,b) \subset U_{I(b)-1}$ for $b > 0$.

2.8 Theorem. The theory of locally pure and torsionfree groups is decidable.

Proof. The theory of linearly ordered groups is decidable [9]. Since every L_m-sentence about the order topology can be translated into an $L'_{\omega\omega}$-sentence about the order (where $L' = L \cup \{<\}$), the L_m-theory of all groups with order topology is decidable. Whence, by 2.7, the theory of all hausdorff, locally pure and torsionfree groups is decidable.

On the other hand the theory of torsionfree groups with the indiscrete topology (which essentially is the $L_{\omega\omega}$-theory of torsionfree groups) is decidable [7].

From these two facts, we obtain the decidability of the theory of locally pure and torsionfree groups using 2.6 and the results on direct sums of I § 6.

2.9 Corollary. The L_m-theory of the group of rationals with its natural topology is axiomatized by

"torsionfree, divisible, $\neq \{0\}$, hausdorff and locally pure".

Proof. All divisible and ordered groups are $L'_{\omega\omega}$-equivalent [19] (where $L' = L \cup \{<\}$).

2.10 Exercise. Prove the completeness of the axiom system of 2.9 by the following quantifier elimination method: Show by induction on φ, that every L_m-formula $\varphi(x_1,\ldots,x_n,X_1,\ldots,X_m)$ is "equivalent" to a quantifierfree formula $\psi(x_1,\ldots,x_n,X_1,\ldots,X_m)$ in the sense that for all models (A,ν) of our axioms, any $a_1,\ldots,a_n \in A$ and any divisible subgroups $A_1,\ldots,A_m \in \nu$ with $A \supsetneqq A_1 \supsetneqq \ldots \supsetneqq A_m$, we have

$$(A,\nu) \models (\varphi \leftrightarrow \psi)\,[a_1,\ldots,a_n,A_1,\ldots,A_m]$$

(use ω_1-saturated (A,ν)).

2.11 Theorem. The theory of algebraically complete groups is decidable.

Proof. By 2.6 and the results of I § 6, it is enough to prove the following lemma.

2.12 Lemma. (A,ν) is a hausdorff, algebraically complete group iff it is L_m-equivalent to a direct sum $\bigoplus_{i \in I}(B_i,\delta_i)$ of groups with the discrete topology.

Proof. By 2.5, one direction is trivial. – Now let (A,ν) be hausdorff and algebraically complete. Choose an ω_1-saturated structure (C,λ) with $(C,\lambda) \equiv_{L_m} (A,\nu)$. Then λ has a basis γ consisting of pure subgroups. – Fix $c \in C$ and $V \in \gamma$. c is contained in a denumerable pure subgroup P of C (cf.[8]). Since nC is closed for $n \in \omega$ and λ is closed under countable intersections, we find a sufficiently small $U \in \gamma$ s.t. $U \subset V$, $U \cap P = \{0\}$, and such that for all $n \in \omega$ and $b \in P$

$$b \notin nC \text{ implies } (b + U) \cap nC = \emptyset .$$

Then the direct sum $U + P$ is pure in C. Let π be the set of all pure subgroups of C. We just have proved:

(*) For all $c \in C$ and $V \in \gamma$ there are $P \in \pi$ and $U \in \gamma$ s.t. $c \in P$, $U \subset V$, $U \cap P = \{0\}$ and $U + P \in \pi$.

Now let (C^*,γ^*,π^*) be an ω_1-saturated three-sorted structure with $(C^*,\gamma^*,\pi^*) \equiv_{L_2} (C,\gamma,\pi)$. All groups in $\gamma^* \cup \pi^*$ are pure. (*) holds in (C^*,γ^*,π^*), if we replace C,γ,π by C^*,γ^*,π^*. But for $P \in \pi^*$ and $U \in \gamma^*$, $P + U$ is again an ω_1-saturated group, and hence a direct factor of C^*, if $P + U \in \pi^*$. Thus, if ϱ denotes the set of all subgroups of C^*, we have:

(**) For all $c \in C^*$ and $V \in \gamma^*$ there are $Q \in \varrho$ and $U \in \gamma^*$ s.t. $c \in Q$, $U \subset V$ and $C^* = U \oplus Q$.

Finally, choose a denumerable (B,β,σ) with $(B,\beta,\sigma) \equiv_{L_2} (C^*,\gamma^*,\varrho)$. Then $(B,\beta) \equiv_{L_m} (A,\nu)$, β and σ consist of subgroups and (**) holds for B,β,σ. Suppose $B = \{b_i | i \in \omega\}$ and $\beta = \{V_i | i \in \omega\}$. We construct, by induction, a basis $(U_i)_{i \in \omega}$ of $\hat{\beta}$ with $U_i \in \beta$ and $B = U_o \supset U_1 \supset U_2 \supset \ldots$ and complements B_i, $U_i = U_{i+1} \oplus B_i$. Suppose U_i has already been defined and let c be the U_i-component of b_i in the decomposition $C = U_i \oplus B_{i-1} \oplus \ldots \oplus B_o$. Using (**) we find subgroups $Q \in \sigma$ and $U_{i+1} \in \beta$ s.t. $c \in Q$, $U_{i+1} \subset U_i \cap V_i$ and $B = U_{i+1} \oplus Q$. Set $B_i = Q \cap U_i$. Then $U_i = B_i \oplus U_{i+1}$, $c \in B_i$ and $b_i \in B_i \oplus \ldots \oplus B_o$. In particular, we have $U_i = B_i \oplus B_{i+1} \oplus B_{i+2} \oplus \ldots$. Hence $(B,\hat{\beta}) = \bigoplus_{i \in \omega} (B_i,\delta_i)$, where δ_i is the discrete topology on B_i. This completes the proof of 2.11 and 2.12.

To prove our last theorem we need the following lemma.

2.13 <u>Lemma</u>. Suppose B is a denumerable group. Let H be the direct sum of countably many non-trivial torsionfree groups. Then there is a hausdorff topology on $B \oplus H$ with respect to which H is a dense subset.

<u>Proof</u>. Let $\{b_n | n \in \omega\}$ be an enumeration of B, and suppose that $H = \bigoplus_{n,i \in \omega} G_{n,i}$ where $G_{n,i}$ are non-trivial and torsionfree. Take $g_{n,i} \in G_{n,i} \setminus \{0\}$. Define the subgroup U_i of $B \oplus H$ by

$$U_i = \text{subgroup generated by } \{b_n - g_{n,j} | j \geq i, n \in \omega\}.$$

Then $\{U_i | i \in \omega\}$ is the basis of a topology with the desired properties.

2.14 <u>Remark</u>. One can prove that for all countable subgroups C of a given group A the following are equivalent.

(i) There is a hausdorff topology on A s.t. C is a dense subset of A.

(ii) Every decomposition $C = \bigcup_{i=1}^{k}(c_i + C[n_i])$ (where $C[n] = \{c \mid nc = 0\}$

yields a decomposition $A = \bigcup_{i=1}^{k}(c_i + A[n_i])$.

2.15 Theorem. The theory of hausdorff topological groups is hereditarily undecidable.

Proof. Let p be a prime number and $q = p^g$. By [2] the theory of all pairs (A,B), where B is a subgroup of the group A and $q \cdot A = \{0\}$, is hereditarily undecidable. We show that in case A is denumerable there is a hausdorff topological group (C,μ) s.t.

$$(A,B) \simeq (C/_{qC}, \overline{qC}/_{qC}) .$$

Then the theory of those pairs is interpretable over the theory of hausdorff topological groups; thus this theory is hereditarily undecidable too.

So, let (A,B) with denumerable A be given and let H be the direct sum of countable many copies of $\mathbb{Q}$. There is a hausdorff topology ν on $B \oplus H$ with respect to which H is a dense subset (see 2.13). Use ν as basis of hausdorff topology μ of $C = A \oplus H$. Then, with respect to μ, $\bar{H} = B \oplus H$. We have $qC = H$, $C/_H \simeq A$ and $\bar{H}/_H \simeq B$.

§ 3 Topological fields

This section consists of three parts:

In part A we characterize several L_t-elementary classes $\mathfrak{K}$ of fields with a topology (e.g. locally bounded fields) by theorems of the following kind:

"$(K,\tau) \in \mathfrak{K}$ iff (K,τ) is L_t-equivalent to a field with a topology given in certain simple manner (e.g. by a subring of K)".

In part B we consider fields with a topology given by a valuation ring (or an ordering). We introduce L_t-axiom systems T s.t.

"$(K,\tau) \models T$ iff (K,τ) is L_t-equivalent to a field with a valuation topology (an order topology)".

For valuation rings, T will be the theory of V-topological fields. We give two applications.

Finally, in part C, we determine the L_t-theory of the field of real numbers and the field of complex numbers with their natural topology.

A. Characterization of topological fields.

Let K $(=(K,+,-,\cdot,0,1))$ be a field and τ a topology on K.

τ is a ring topology, if τ is hausdorff, non-discrete and $+,-,\cdot$ are continuous. As in the case of topological abelian groups, τ is determined by the filter ν of neighborhoods of 0.

We identify (K,τ) and (K,ν), and use L_m for (K,ν) instead of L_t for (K,τ).

Let ν be a monotone system on the field K. ν is a ring topology, if the following L_m-sentences hold in (K,ν):

(0) $\forall X\ \forall Y\ \exists Z\ Z \subset X \cap Y$

(1) $\forall x \neq 0\ \exists X\ x \notin X$

(2) $\forall X\ \ \{0\} \subsetneqq X$

(3) $\forall X\ \exists Y\ \ Y - Y \subset X$

(4) $\forall X\ \exists Y\ \ Y \cdot Y \subset X$

(5) $\forall X\ \forall x\ \exists Y\ \ xY \subset X.$

A ring topology ν is a field topology (and (K,ν) is a topological field), if

$x \to x^{-1}$ is continuous. This means that in addition to (0) – (5) the following axiom (6) holds:

(6) $\forall X \; \exists Y \quad (1+Y)^{-1} \subset 1 + X$.

A subset S of K is <u>bounded</u> (w.r.t.ν), if for every $U \in \nu$ there is a $V \in \nu$ s.t. $V \cdot S \subset U$. A ring topology ν is <u>locally bounded</u>, if ν contains a bounded set. This can be expressed by the L_m-sentence:

(7) $\exists X \; \forall Y \; \exists Z \quad Z \cdot X \subset Y$.

Let R be a proper subring of K (understood to contain 1) s.t. K is the quotient field Quot(R) of R. As it is easily seen, the set of all non-zero ideals of R is a locally bounded ring topology ν_R of K (ν_R is hausdorff, since R is not a field; (5) holds because K = Quot(R); R is bounded). We call ν_R a <u>standard locally bounded topology</u>.

If in addition R is a local ring (i.e. R only has one maximal ideal M), ν_R is a field topology. ((6) holds because $(1+xM)^{-1} \subset (1+xM)$ for all $x \in R$). In this case we call ν_R a <u>standard locally bounded field topology</u>.

Let ν be a filter which is the union of standard locally bounded (field) topologies. Then ν is a ring (field) topology. We call such a ν a <u>standard (field) topology</u>.

3.1 <u>Lemma</u>. Let K be a field and ν a monotone system on K which is closed under countable intersections. If ν is a (locally bounded) ring (field) topology, then ν is a standard (locally bounded) (field) topology.

<u>Proof</u>. Choose an enumeration $\{f_i | i \in \mathbb{N}\}$ of the prime field F of K.- Let $U_o \in \nu$ be given. Using (1), (3), (4), (5) we construct, by induction, a sequence $U_o \supset U_1 \supset U_2 \supset \ldots$ of elements of ν s.t.

$$-1 \notin U_1,\; U_{i+1} - U_{i+1} \subset U_i,\; U_{i+1} \cdot U_{i+1} \subset U_i, \{f_o, f_1, \ldots, f_i\} \cdot U_{i+1} \subset U_i .$$

(In the case that ν is a field topology we choose – using (6) – the sequence in such a way that in addition $(1+U_{i+1})^{-1} \subset 1 + U_i$).

The intersection $J = \bigcap_{i \in \omega} U_i$ belongs to ν and satisfies

$-1 \notin J$, $J - J \subset J$, $J \cdot J \subset J$, $F \cdot J \subset J$ (and $(1+J)^{-1} \subset 1 + J$ in the case of a field topology).

$R = F + J$ is a proper subring of K and - by (2) - J is a non-zero ideal of R. (If $(1 + J)^{-1} \subset 1 + J$ and $f + j \in R \setminus J$, then $(f + j)^{-1} = f^{-1}(1 + f^{-1}j)^{-1} \in F \cdot (1 + J)^{-1} \subset R$. This shows that $R \setminus J$ consists of units, i.e. R is local).

We show that $\mathrm{Quot}(R) = K$. Let b be an arbitrary element of K. Choose $V \in \nu$ with $bV \subset R$ and $c \in (V \cap R) \setminus \{0\}$. Then $b = \frac{a}{c}$, where $a = bc \in R$.

Thus ν_R is a standard locally bounded (field) topology. Since every $V \in \nu_R$ contains a principal ideal aR, $a \neq 0$, which belongs to ν (by (5)), we have $\nu_R \subset \nu$. - U_o was an arbitrary element of ν. U_o belongs to ν_R and thus we have shown that ν is the union of topologies of the form ν_R, i.e. ν is a standard (field) topology.

Now suppose that ν is locally bounded. We start the above construction with a bounded $U_o \in \nu$. Then $J \in \nu_R$ is bounded (w.r.t.ν). Let $W \in \nu$. There is $V \in \nu$ with $VJ \subset W$. Choose $a \in V \setminus \{0\}$. Then $aJ \subset W$; this shows that $W \in \nu_R$ and $\nu_R = \nu$

3.2 Remark. We have shown that a field topology closed under countable intersections has a basis consisting of maximal ideals of local rings $R \subsetneqq K$ with $\nu_R \subset \nu$ and $\mathrm{Quot}(R) = K$.

3.3 Theorem. Let ν be a monotone system on the field K. ν is a (locally bounded) ring (field) topology iff (K,ν) is L_m-equivalent to (F,μ), where μ is a standard (locally bounded) (field) topology on F.

Proof. This follows from 3.1 and the fact that μ is closed under countable intersections, if μ is a filter (i.e. (0) holds) and (F,μ) is ω_1-saturated (cf. 1.18).

3.4 Exercise. Let K be a field. A condition is a finite set $p(X_1,\dots,X_n)$ of "formulas" of the form $a \in X_i$ $(a \in K)$, $\forall x_1 \dots x_k \in X_i \rightarrow h(x_1,\dots,x_k) \in X_j$ (where h is a polynomial with coefficients in K), which is satisfied by some sets $U_i \in P(K)$ with $0 \in U_i$. We write $\emptyset \Vdash \varphi$ if player I has a winning strategy in the following game: Players I and II by turns choose the elements of a sequence $p_o \subset p_1 \subset p_2 \subset \dots$ of conditions. I wins if $\bigcup_{i \in \mathbb{N}} p_i$ is satisfied by a sequence $(U_i)_{i \in \mathbb{N}}$ s.t. $0 \in U_i$ and $(K, \{U_i | i \in \mathbb{N}\}) \models \varphi$.

Show: a) If K is countable, then there is $\alpha \subset P(K)$ s.t.

for all L_m-sentences φ,

$$(K,\alpha) \models \varphi \quad \text{iff} \quad \emptyset \Vdash \varphi.$$

b) $\emptyset \Vdash$ "ring topology, not locally bounded".

B. Valued and ordered fields.

A subring A of K is a valuation ring, if $A \neq K$ and if $x \in A$ or $x^{-1} \in A$ holds for all $x \in K$. Then we call ν_A a valuation topology.

Let A be a valuation ring of K. $M = \{x | x^{-1} \notin A\}$ is the set of non-units of A. Clearly, $AM \subset M$. Suppose $x,y \in M$ and e.g. $xy^{-1} \in A$. Then $x + y \notin M$ would imply $y^{-1} = (x+y)^{-1}(1 + xy^{-1}) \in A$. We conclude that A is a local ring with maximal ideal M. Therefore ν_A is a field topology.

3.5 Remark. (cf.[4]). A valuation is a surjective map $v\colon K \to \Gamma \cup \{\infty\}$, where Γ is a non-trivial ordered abelian group, which satisfies

$$v(x) = \infty \quad \text{iff} \quad x = 0$$

$$v(xy) = v(x) + v(y) \qquad (\text{where } g + \infty = \infty + g = \infty + \infty = \infty)$$

$$v(x+y) \leq \max\{v(x), v(y)\} \qquad (\text{where } g < \infty)\ .$$

Valuation rings are just the rings of the form $A = \{x | v(x) \geq 0\}$. A determines v up to a natural isomorphism. A basis of ν_A is given by

$$U_g = \{x | v(x) > g\}, \quad \text{where } g \in \Gamma\ .$$

3.6 Theorem. (K,ν) is L_m-equivalent to a field with a valuation topology iff ν is a V-topology, i.e. a ring topology which satisfies

$$(8) \qquad \forall X\, \exists Y\, \forall x\, \forall y \quad (x \notin X \wedge y \notin X \quad \rightarrow \quad xy \notin Y).$$

Proof. Let A be a valuation ring. $\{aM | a \in A\setminus\{0\}\}$ forms a basis of ν_A. Thus the sequence of implications

$$x,y \notin aM \;\Rightarrow\; ax^{-1}, ay^{-1} \in A \;\Rightarrow a^2(xy)^{-1} \in A \;\Rightarrow\; xy \notin a^2M$$

shows that ν_A is a V-topology.

Conversely, assume that ν is a V-topology. Choose (F,μ) L_m-equivalent to (K,ν) s.t. μ is closed under countable intersections. - Let $U_o \in \mu$. Since μ is a V-topology, we can construct a sequence $U_o \supset U_1 \supset U_2 \supset \ldots$ of elements of μ s.t.

$$1 \notin U_1,\ U_{i+1} + U_{i+1} \subset U_i,\ U_{i+1} \cdot U_{i+1} \subset U_i \quad \text{and s.t.}$$

$$x,y \notin U_i \;\text{ implies }\; xy \notin U_{i+1}.$$

The intersection $M = \bigcap_{i \in \omega} U_i$ belongs to μ and satisfies

$$1 \notin M,\ M + M \subset M,\ M \cdot M \subset M,\ x,y \notin M \Rightarrow xy \notin M.$$

Set $A = \{x \mid x^{-1} \notin M\}$.

It is easy to see that $\pm 1 \in A$, $A \neq K$, $AA \subset A$ and $x \in A$ or $x^{-1} \in A$ for all $x \in K$. – Suppose $x,y \in A$. We want to show that $x + y \in A$. We can assume that $xy^{-1} \notin M$ (otherwise $yx^{-1} \notin M$). Since $xy^{-1}(1 - x(x+y)^{-1}) = x(x+y)^{-1}$, $x(x+y)^{-1} \in M$ would imply $(1 - x(x+y)^{-1}) \in M$ and $1 \in M$. Therefore $x(x+y)^{-1} \notin M$ and (as $x^{-1} \notin M$) $(x+y)^{-1} \notin M$, i.e. $x + y \in A$. – Hence A is a valuation ring with maximal ideal $M \in \mu$.

Clearly $\nu_A \subset \mu$. We finish the proof showing that $\bar{\nu} \subset \bar{\mu}$ implies $\bar{\nu} = \bar{\mu}$, whenever $\bar{\nu}$ is a ring topology and $\bar{\mu}$ is a V-topology ("V-topologies are minimal ring topologies").

Let $U \in \bar{\mu}$. Choose $V \in \bar{\nu}$ and $U' \in \bar{\mu}$ s.t. $1 \notin V \cdot V$ and s.t. $x,y \notin U \cap V$ implies $xy \notin U'$. Take $a \in U' \setminus \{0\}$. Then $aV \subset U$ – and $U \in \bar{\nu}$ – for $x \in V$ implies $x^{-1} \notin V$, and hence $ax \in U$.

3.7 <u>Remarks and exercises</u>.

(a) We have shown that every V-topology ν, which is closed under countable intersections has a basis consisting of maximal ideals of valuation rings A with $\nu = \nu_A$.

(b) It is shown in [12] that the fields with V-topology are just the fields with valuation topology and the subfields of $\mathbb{C}$ with their natural topology. This leads to another proof of 3.6, since the topology of a subfield of $\mathbb{C}$ is not closed under countable intersections.

(c) Prove 3.6 using 3.3 and the fact that every subring of K, which is not a field, is contained in a valuation ring of K.

(d) Show: (K,ν) is a V-topological field iff for every topological field (F,μ) L_m-equivalent to (K,ν), μ is a minimal ring topology (use (c)).

As an application we prove two well-known theorems about V-topologies.

3.8 (<u>Approximation theorem</u>). Let $\nu_1,\ldots,\nu_n$ be different V-topological on K. Suppose $V_i \in \nu_i$ and $a_i \in K$ for $i = 1,\ldots,n$. Then there is an element $b \in K$ s.t.

$$b - a_1 \in V_1, \ldots, b - a_n \in V_n .$$

Proof. Let L_m be the monotone language appropriate for structures of type $(K, \nu_1, \ldots, \nu_n)$ and denote by $X_i, Y_i, \ldots$ the variables ranging over ν_i. In this language it is expressible that the ν_i are different V-topologies. We have to show that $(K, \nu_1, \ldots, \nu_m)$ is a model of the following L_m-sentence

$$\forall X_1 \forall X_2 \ldots \forall X_n \forall x_1 \ldots \forall x_n \exists y (y - x_1 \in X_1 \wedge \ldots \wedge y - x_n \in X_n).$$

We can assume that all ν_i are closed under countable intersections and that $\nu_i = \nu_{A_i}$ holds for some valuation rings A_i.

We quote the approximation theorem for valuation rings

> Suppose $B_1, \ldots, B_n$ are pairwise independent valuation rings, $N_1, \ldots, N_n$ non-zero ideals of $B_1, \ldots, B_n$ and $b_1, \ldots, b_n \in K$. Then there is an element $b \in K$ s.t. $b - a_1 \in N_1, \ldots, b - a_n \in N_n$.

Thus, it remains to show that the A_i are independent, i.e. that no valuation ring B contains two of the A_i. But $A_i, A_j \subset B$ implies $\nu_B \subset \nu_{A_i}, \nu_{A_j}$. Hence, $\nu_B = \nu_{A_i} = \nu_{A_j}$ and therefore $i = j$.

3.9 Continuity of roots. Let ν be a V-topology on K, $U \in \nu$ and $a_1, \ldots, a_n \in K$. Then there is $V \in \nu$ s.t. for all $b_1, \ldots, b_n$

$$\prod_{i=1}^{n}(X - a_i) - \prod_{i=1}^{n}(X - b_i) \in V[X] \text{ implies } a_i - b_{\pi(i)} \in U, 1 \le i \le n,$$

for a permutation π of $1, \ldots, n$.

(where $V[X]$ denotes the set of polynomials with coefficients in V).

Proof. For every n we have to show that (K, ν) satisfies a certain L_m-statement. We can assume that ν is closed under countable intersections. We find a valuation ring A with maximal ideal $M \in \nu$ s.t. $M \subset U \setminus \{a_1^{-1}, \ldots, a_n^{-1}\}$. Then $\prod_{i=1}^{n}(X - a_i) = f \in A[X]$. We set $V = M$. $f - \prod_{i=1}^{n}(X - b_i) \in M[X]$ implies $g = \prod_{i=1}^{n}(X - b_i) \in A[X]$ and $b_1, \ldots, b_n \in A$, since A is integrally closed. We have $g \equiv f \pmod M$ and thus $a_1 \equiv b_{\pi(1)}, \ldots, a_n \equiv b_{\pi(n)} \pmod M$ for some permutation π.

An ordering of K is a linear order $<$ satisfying the axioms

$$\forall x \forall y (x < y \rightarrow x + z < y + z) \text{ and } \forall x \forall y (0 < x \wedge 0 < y \rightarrow 0 < xy).$$

Denote by $\nu_<$ the monotone system generated by the open intervals $(-a,a) = \{x \mid -a < x < a\}$ where $0 < a$. $\nu_<$ is a topology. We call $\nu_<$ an <u>order topology</u>.

3.10 <u>Remark</u>. Orderings correspond to "positive cones" P via the definitions $P = \{x \mid 0 \leq x\}$ resp. $x \leq y$ iff $y - x \in P$. By definition, P is a positive cone iff $P + P \subset P$, $PP \subset P$, $P \cap -P = \{0\}$ and $P \cup -P = K$.

3.11 <u>Theorem</u>. A topological field (K,ν) is L_m-equivalent to a field with order topology iff ν is a V-topology which satisfies for every $n \in \mathbb{N}$

(9) $$\exists X \ \forall x_1 \ldots \forall x_n \ \ x_1^2 + \ldots + x_n^2 \notin -1 + X.$$

<u>Proof</u>. Taking as X the set $(-1,1)$ we see that an order topology satisfies (9). Conversely, assume that ν is a V-topology and that (9) holds for every n. Choose (F,μ) L_m-equivalent to (K,ν) with μ closed under countable intersections. For every n there is a $U_n \in \mu$ s.t. $-1 + U_n$ does not contain a sum of n squares. By 3.7 (a) there is a valuation ring A with maximal ideal M s.t. $\mu = \nu_A$ and $M \subset \bigcap_{i \in \mathbb{N}} U_i$.

Set $Q = \{\sum_{i=1}^{n} (1 + m_i)x_i^2 \mid n \in \mathbb{N},\ m_i \in M,\ x_i \in K\}$. Then $1 + M \subset Q$, $K^2 \subset Q$ and

a) $Q + Q \subset Q$, $Q \cdot Q \subset Q$

b) $-1 \notin Q$.

To see b), suppose $-1 = \sum_{i=1}^{n} (1 + m_i)x_i^2$. We can assume that $\frac{x_i}{x_1} \in A$ for $i = 1,\ldots,n$. Then

$$x_1^{-2} + \sum_{i=2}^{n} \left(\frac{x_i}{x_1}\right)^2 = -1 - \sum_{i=1}^{n} m_i \left(\frac{x_i}{x_1}\right)^2 \in -1 + M,$$

which contradicts the choice of M.

Now let P be a maximal extension of Q which satisfies a) and b). We show that P is a positive cone. Since $\frac{1}{p^2} \in P$, $p \in P \cap -P$ implies $\frac{1}{p} \in P$ and $-1 = -p \cdot \frac{1}{p} \in P$. Whence $P \cap -P = \{0\}$. Let $x \in K$. Since $x^2 \in P$ the sets $Q_1 = P + xP$ and $Q_2 = P - xP$ satisfy a). Q_1 or Q_2 also must satisfy b). For otherwise we have $-1 = p_1 + xq_1 = p_2 - xq_2$ with $p_i, q_i \in P$. This implies $-(q_1 + q_2) = p_1q_2 + p_2q_1$ hence $q_1 + q_2 = 0$, which implies $q_1 = q_2 = 0$, and $-1 = p_1$, a contradiction. Thus $P = Q_1$ or $P = Q_2$, i.e. $x \in P$ or $x \in -P$. We have shown that $K = P \cup -P$.

By ($x \leq y$ iff $y - x \in P$) we obtain an ordering of K with $M \subset (-1,1)$. $\nu_< \subset \nu_A$ implies $\nu_< = \mu$.

3.12 Remarks (see [15],[16]).

(a) There are V-topologies which satisfy (9) but are not order topologies.

(b) A PC-class $\mathfrak{K}$ of monotone structures is the class of L-reducts of an L'_m-theory T where $L \subset L'$. – Note that then $(\mathfrak{A},\nu) \models Th(\mathfrak{K})$ iff $(\mathfrak{A},\nu)$ is L_m-equivalent to some $(\mathfrak{B},\mu) \in \mathfrak{K}$.

If T is recursively enumerable, then $Th(\mathfrak{K})$ (which is $\{\varphi \in L_m | T \models \varphi\}$) also is recursively enumerable. The classes of fields with standard (locally bounded field) topology, non-minimal ring topology, valuation topology and order topology are PC-classes. We gave explicit axiomatizations of the respective $Th(\mathfrak{K})$. –

For a lot of natural classes $\mathfrak{K}$ of V-topological fields, $Th(\mathfrak{K})$ is not recursively enumerable. E.g. the class of subfields of $\mathbb{C}$ or the class of fields with a topology induced by an absolute value.

C. Real and complex numbers.

Let ϱ be the natural topology of the field $\mathbb{R}$ of real numbers. The fields elementarily equivalent to $\mathbb{R}$ are just the real closed fields. Since the (unique) ordering of real closed fields and henceforth their order topology are elementarily definable, the fields (K,ν) L_m-equivalent to $(\mathbb{R},\varrho)$ are just the real closed fields with order topology ν. This class is axiomatizable by a set of L_m-sentences. We give another description of this class.

3.13 Theorem. (K,ν) is L_m equivalent to $(\mathbb{R},\varrho)$ iff K is real closed, ν is a V-topology, and

(10) $$(K,\nu) \models \exists X\ \forall x\ x^2 \notin -1 + X.$$

Proof. One direction is trivial. On the other hand let K be real closed, let ν be a V-topology and suppose (10) holds. Since sums of squares are squares, (9) holds and by 3.11, (K,ν) is L_m-equivalent to $(F,\nu_<)$, where $<$ is an ordering of F. F is real closed, thus $<$ is the unique ordering of F. Hence, $(F,<)$ and $(\mathbb{R},<)$ are elementarily equivalent. This implies

$(F,\nu_<) \equiv_{L_m} (R,\nu_<) = (R,\varrho)$.

Finally, we give an axiomatization of the L_m-theory of $(\mathbb{C},\gamma)$, where γ is the natural topology of the field $\mathbb{C}$ of complex numbers.

3.14 <u>Theorem</u>. (K,ν) is L_m-equivalent to $(\mathbb{C},\gamma)$ iff K is an algebraically closed field of characteristic 0 and ν is a V-topology.

<u>Proof</u>. Since γ is a V-topology, one direction is clear. For the converse we first note that every (K,ν), where K is algebraically closed and of characteristic 0, and where ν is a V-topology, is L_m-equivalent to a field (F,ν_A), where A is a valuation ring with residue class field $^A/M$ of characteristic 0. To obtain this result take (F,μ) L_m-equivalent to (K,ν) with μ closed under countable intersections. There is $U \in \mu$ s.t. $U \cap \mathbb{N} = \{0\}$. Now choose A s.t. $\mu = \nu_A$ and $M \subset U$.

By a result of Robinson [ ] all structures (F,A) of the above form are elementarily equivalent. Since ν_A is definable in terms of A, any two (F,ν_A) are L_m-equivalent. Hence, any two (K,ν) with K algebraically closed of characteristic 0 and ν a V-topology are L_m-equivalent.

3.15 <u>Corollary</u>. The L_m-theory of $(\mathbb{C},\gamma)$ is decidable.

3.16 <u>Exercise</u>. Derive the decidability of $(\mathbb{C},\gamma)$ by an interpretation in $(R,<)$.

§ 4 Topological vector spaces.

A topological vector space is a two-sorted topological structure

$$((K,\nu),(V,\mu))$$

(more precisely $(((K,+,-,\cdot,0,1),\nu),((V,+,-,0,),\mu),\circ))$ where the first structure is a topological field, the second structure is a hausdorff topological abelian group, V is a vector space over K, and where the scalar multiplication $\circ : K \times V \to V$ is continuous. ν and μ are the neighborhood filters of the corresponding zero elements.

Let L_m denote the monotone language appropriate for topological vector spaces. We use the following symbols as variables

$\alpha,\beta,\ldots$ for variables ranging over K,

$A,B,\ldots$ for variables ranging over ν,

$x,y,\ldots$ for variables ranging over V,

$X,Y,\ldots$ for variables ranging over μ.

Thus, given K and V and monotone systems ν and μ on K resp. V, $((K,\nu),(V,\mu))$ is a topological vector space iff (K,ν) is a topological field, (V,μ) is a hausdorff topological abelian group, V is a vector space over K and $((K,\nu),(V,\mu))$ is a model of

(1) $\forall X\ \forall x\ \exists A \quad Ax \subset X.$

(2) $\forall X\ \forall \alpha\ \exists Y \quad \alpha Y \subset X.$

(3) $\forall X\ \exists A\ \exists Y \quad AY \subset X.$

This shows that the class of topological vector spaces can be axiomatized by finitely many L_m-sentences.

Let $((K,\nu),(V,\mu))$ be a topological vector space. A subset S of V is bounded, if for every $U \in \mu$ there is a $W \in \nu$ s.t. $WS \subset U$. – $((K,\nu),(V,\mu))$ is locally bounded, if μ contains a bounded element, i.e. if $((K,\nu),(V,\mu))$ is a model of

(4) $\exists X\ \forall Y\ \exists A \quad AX \subset Y.$

Let T_{LB} denote the theory of locally bounded vector spaces. We have no results for non-bounded vector spaces. In this section we consider four types of structures (which occur in mathematics) and describe their topological properties:

locally bounded real vector spaces,
locally bounded real vector spaces with a distinguished subspace,
Banach spaces with linear mappings,
dual pairs of normal spaces.

A. Locally bounded real vector spaces.

We introduce the theory T_R axiomatized by

T_{LB}, "K is a real closed field with its order topology (cf. 3.13)" ,

and for $n = 0,1,2...$ the axiom

(5) $$\forall X \, \exists Y \, \forall x_1 \ldots \forall x_n \, \forall y \notin \langle x_1,\ldots,x_n\rangle \, \exists x \, (x \in X \cap \langle x_1,\ldots,x_n,y\rangle \wedge (x + Y \cap \langle x_1,\ldots,x_n\rangle) = \emptyset) \quad .$$

Here $\langle x_1,\ldots,x_n\rangle$ denotes the subspace spanned by $x_1,\ldots,x_n$.

4.1 Theorem. a) Every locally bounded real vector space is a model of T_R.
b) Two models of T_R are L_m-equivalent iff they have the same dimension.

4.2 Remark. We do not distinguish infinite dimensions. Any infinite dimensional vector space has dimension ∞.

Proof of 4.1 a): Let (V,μ) be a locally bounded real vector space. We have to show that (5) holds. In real vector spaces, finite dimensional subspaces are closed. Thus we are done, if for every $U_1 \in \mu$ we can exhibit an $U_2 \in \mu$ s.t.

for any closed subspace $H \subset V$, and any $v \notin H$ there is
$u \in U_1 \cap (H + \langle v\rangle)$ s.t. $(u + U_2) \cap H = \emptyset$.

Given U_1 we choose a bounded open U_2 with $-2U_2 \subset U_1$.
Put $b = \sup\{\alpha \mid (-\alpha U_2) \cap (v + H) = \emptyset\}$. Then, we have $0 < b$ (since $0 \notin v + H$, $v + H$ is closed and U_2 is bounded), $b < \infty$ (by (1)) and $(-bU_2) \cap (v + H) = \emptyset$ (since U_2 is open). There is $w \in (-2bU_2) \cap (v + H)$. Set $u = b^{-1}w$. Then $u \in -2U_2 \subset U_1$ and $(u + U_2) \cap H = \emptyset$ (since $(-bU_2) \cap (w + H) = \emptyset$).

4.3 Remark. For normed vector spaces (5) is an immediate consequence of Riesz'lemma.

Riesz'lemma : Given a closed subspace H, $v \in H$ and $a > 0$ there is is $u \in H + \langle v\rangle$ s.t. $\|u\| = 1$ and $\inf\{\|u - u'\| \mid u' \in H\} > 1 - a$.

For the proof of b) we determine the saturated models of T_R.

4.4 <u>Definition</u>. An <u>euclidean vector space</u> is a vector space V over a real closed field K together with a euclidean (= positive definite symmetric bilinear) form (,). For $a \in K$, $a > 0$, denote by B_a the ball

$$B_a = \{v \in V \mid \|v\| \le a\} \qquad (\text{where } \|v\| = \sqrt{(v,v)}).$$

$\{B_a \mid a > 0\}$ is the basis of a monotone system μ. (V,μ) is a locally bounded topological vector space. We call (V,μ) a <u>euclidean topological vector space</u>.

4.5 <u>Lemma</u>. $((K,\nu),(V,\mu))$ is a model of T_R iff it is L_m-equivalent to a euclidean topological vector space.

<u>Proof. of 4.1 b)</u>: Let $(K_i,V_i,(\, , \,))$, $i = 1,2$ be euclidean vector spaces of the same dimension. Denote by ν_i resp. μ_i the corresponding topologies on K_i resp. V_i. Since K_1 and K_2 are elementarily equivalent there are euclidean vector spaces $(K,V_i',(\, , \,))$, $i = 1,2$, over the same field K s.t.

$$(K_i,V_i,(\, , \,)) \equiv (K,V_i',(\, , \,)) \qquad i = 1,2 .$$

Furthermore we can assume that the $(K,V_i',(\, , \,))$ are denumerable. But then, the $(K,V_i'(\, , \,))$ have orthonormal bases and hence, are isomorphic being of the same dimension. This shows

$$(K_1,V_1,(\, , \,) \equiv (K_2,V_2,(\, , \,)),$$

which implies $((K_1,\nu_1),(V_1,\mu_1)) \equiv_{L_m} ((K_2,\nu_2),(V_2,\mu_2))$.

<u>Proof of 4.5</u>: First we show that any euclidean topological vector space $((K,\nu,),(V,\mu))$ is a model of T_R . Note that any finite dimensional subspace F of V gives rise to an orthogonal decomposition

$$V = F \oplus F^{\perp}$$

where $F^{\perp} = \{x \mid (x,y) = 0 \text{ for all } y \in F\}$.

To show (5) suppose that $U_1 = B_a$ is given. Put $U_2 = \{x \mid \|x\| < a\}$. Then, for $v \notin F$ we only have to choose $u \in F^{\perp} \cap (F + \langle v\rangle)$ of lenghth a to get

$$u \in U_1 \quad \text{and} \quad (u + U_2) \cap F = \emptyset .$$

Now assume that $((K,\nu,),(V,\mu))$ is a model of T_R. We proceed to an ω_1-saturated L_m-equivalent structure $((K_1,\nu_1),(V_1\mu_1))$. Since μ_1 is closed under countable intersections we find a bounded $U_1 \in \mu_1$ and $\tilde{U} \subset U_1$ s.t. for any finite dimensional F and any $v \notin F$ there is $u \in U_1 \cap (F + \langle v\rangle)$ with $(u + \tilde{U}) \cap F = \emptyset$. There is $M_1 \in \nu_1$ with $M_1U_1 + M_1U_1 + \ldots \subset \tilde{U}$.

Since ν_1 is closed under countable intersections, we can assume that M_1 is the maximal ideal of a valuation ring (3.7 a)).

Let $(((K_2,M_2),\nu_2),((V_2,U_2,D_2),\mu_2))$ be denumerable and L_m^*-equivalent to $(((K_1,M_1),\nu_1),((V_1,U_1,D_1),\mu_2))$ for the corresponding L^*, where $D_1 = M_1U_1 + M_1U_1 + \ldots$. $((K_2,\nu_2),V_2,\mu_2))$ will turn out to be euclidean.

Let $v_1,v_2,\ldots$ be a basis of V_2. Set $F_i = \langle v_1,\ldots,v_i\rangle$. We choose $u_i \in U_2 \cap F_i$ s.t. $(u_i + D_2) \cap F_{i-1} = \emptyset$. Then $u_1,u_2,\ldots$ is a basis of V_2, and we have for any $a_1,\ldots,a_n \in K$.

$$a_1,\ldots,a_n \in M_2 \quad \text{iff} \quad a_1u_1+\ldots+a_nu_n \in D_2.$$

For, $a_1,\ldots,a_n \in M_2$ implies $a_1u_1+\ldots+a_nu_n \in M_2U_2+\ldots+M_2U_2 \subset D_2$.

For the converse note that $F_i = \langle u_1,\ldots,u_i\rangle$. Hence, if $a_1u_1+\ldots+a_nu_n \in D_2$, then we obtain step by step

$$(u_n + a_n^{-1}D_2) \cap F_{n-1} \neq \emptyset;\ a_n^{-1}D_2 \not\subset D_2;\ a_n^{-1} \notin A_2;\ a_n \in M_2$$

(since M_2 is the maximal ideal of a valuation ring A_2);$a_1u_1+\ldots a_{n-1}u_{n-1} \in D_2$ etc. .

Now define the euclidean form (,) in such a way that $(u_i)_{i \in \omega}$ is an orthonormal basis. We finish the proof showing that (,) induces μ_2. Choose $a,b \in K_2$, $a,b > 0$, s.t. $[-a,a] \subset M_2 \subset [-b,b]$. Given $U \in \mu_2$ we have $cD_2 \subset U$ for some $c > 0$. Then $B_{ca} \subset U$, for $a_1u_1+\ldots+a_nu_n \in B_{ca}$ implies $a_i \in [-ca,ca]$. Conversely, if $c > 0$ is given, then we have $cb^{-1}D_2 \subset B_c$, for $a_1,\ldots,a_n \in M_2$ implies $\sqrt{a_1^2+\ldots+a_n^2} \in M_2$.

4.6 <u>Remark</u>. Note that, by the same proof, we obtain for any model $((K,\nu),(V,\mu),P_1,\ldots,P_n)$ of T_R with additional predicates $P_1,\ldots,P_n$ an $(L \cup \{P_1,\ldots,P_n\})_m$ - equivalent denumerable structure $((K',\nu'),(V',\mu'),P_1',\ldots,P_n')$ where $((K',\nu'),(V',\mu')$ is an euclidean vector space.

4.7 <u>Corollary</u>. T_R is decidable. The L_m-theory of any locally bounded real vector space is decidable.

4.8 Exercise. Let T_V be the theory T_R where the axioms "K is real closed with its order topology" is replaced by "K is a field with a V-topology". Show:

a) Every locally bounded vector space over a complete field with an absolute value is a model of T_V.
($| \, |: K \to \mathbb{R}_{\geq 0}$ is an absolute value, if for any $a, b \in K$, $|a| = 0$ iff $a = 0$; $|a + b| \leq |a| + |b|$; $|ab| = |a||b|$ and if $| \, |$ takes arbitrarily small positive values ([4])).

b) Two models $((K_i,\nu_i),(V_i,\mu_i))$, $i = 1,2$, of T_V are L_m-equivalent iff $K_1 \equiv K_2$ and V_1 and V_2 have the same dimension.

c) A structure is a model of T_V iff it is L_m-equivalent to a structure of the form

$$((K,\nu),(\bigoplus_{i \in I} K,\mu))$$

where (K,ν) is a V-topological field and $\{B_W | W \in \nu\}$ is a basis of μ where $B_W = \bigoplus_{i \in I} W$.

Hint: a), b) and one direction of c) can be proved like 4.1 a), b) and the corresponding direction of 4.5. It remains to show that the structures described in (c) satisfy (5):

This is clear in case $K = \mathbb{R}$ or $K = \mathbb{C}$ (by the Riesz lemma for (complex) normal vector spaces). From this, the validity of (5) follows for all subfields of $\mathbb{C}$. It remains the case that ν is a valuation topology ν_A of a valuation ring A of K (cf. [12]). Then (5) holds in the form:

For any finite dimensional $F \subset V$ and any $v \notin F$ there is
$u \in B_A \cap (F + \langle v \rangle)$ s.t. $(u + B_M) \cap F = \emptyset$,

(where M denotes the maximal ideal of A).

This is easily shown in case that F is generated by "axis vectors" $(a_i)_{i \in I}$ where $a_i \neq 0$ for exactly one $i \in I$. The general case can be reduced to this situation, for the "Elementarteilersatz" for valuation rings implies that every finite sequence of linearly independent vectors can be mapped to a sequence of axis vectors by an K-automorphism of V which preserves B_A (and hence B_M). (cf. [3],[4].)

B <u>Locally bounded real vector spaces with a distinguished subspace.</u>

In this part we look at structures

$$((K,\nu),((V,H),\mu))$$

where H is a subspace of the topological vector space $((K,\nu),(V,\mu))$. Thus we add to L a unary predicate symbol P. Put $L' = L \cup \{P\}$.

Let T_C be the theory axiomatized by

T_R ,"P is a subspace"

and the L'_m-sentences

(6) $\forall X\ \exists Y\ \forall x_1 \ldots \forall x_n\ \forall y \notin P + \langle x_1,\ldots,x_n\rangle \exists x$

$$(x \in X \cap (P + \langle x_1,\ldots,x_n,y\rangle \wedge (x + Y) \cap (P + \langle x_1,\ldots,x_n\rangle) = \emptyset$$

where $n = 0,1,\ldots$).

4.9 Theorem. a) Every locally bounded real vector space with a distinguished closed subspace is a model of T_C.

b) Two models $((K_i,\nu_i),((V_i,H_i),\mu_i))$, $i = 1,2$, of T_C are L'_m-equivalent iff

$$\dim H_1 = \dim H_2 \quad \text{and} \quad \dim {}^{V_1}/H_1 = \dim {}^{V_2}/H_2$$

(where dim... denotes the dimension of ...).

Proof. The proof of a) is similar to that of 4.1 a) and uses the fact that in real topological vector spaces, $H + \langle v_1,\ldots,v_n\rangle$ is closed whenever H is closed.

Part b) is obtained from the next lemma in a similar way as 4.1 b) from 4.5.

4.10 Lemma. $((K,\nu),((V,H),\mu))$ is a model of T_C iff it is L'_m-equivalent to a euclidean topological vector space with a distinguished subspace which has an orthogonal complement w.r.t. the given euclidean form. (A similar remark as 4.6 applies.)

Proof. For one direction argue as in the proof of 4.5 and note that $H + \langle v_1,\ldots,v_n\rangle$ has an orthogonal complement whenever H has an orthonogal complement.

Let $((K,\nu),((V,H),\mu))$ be a model of T_C. By (6), H is closed and the quotient space $((K,\nu),({}^V/H,{}^{\mu}/H))$ is a model of T_R. Let $f:V \to {}^V/H$ be the natural projection. Applying twice 4.5 (and 4.6) we obtain a denumerable model $((K_1,\nu_1),((V_1,H_1),\mu_1),(V_2,\mu_2),f_1)$ L^*_m-equivalent to $((K,\nu),((V,H),\mu),({}^V/H,{}^{\mu}/H),f)$ (for suitable L*), where μ_1 and μ_2 are induced by euclidean forms $(\ ,\ )^1$ and $(\ ,\ )^2$. f_1 is an open map, thus there is an $\varepsilon > 0$ s.t.

$$B_1^2 \subset f(B_\varepsilon^1)$$

(where B_δ^i is the δ-ball defined by $(\ ,\ )^i$).

Let $(u_i)_{i<k}$ where $k \leq \omega$ be an orthonormal basis of V_2. Choose $v_i \in f_1^{-1}(u_i) \cap B_\varepsilon^1$. Then the linear map $g:V_2 \to V_1$ with $g(u_i) = v_i$ is continuous and $f\circ g = \mathrm{id}$. Thus (V_1,μ_1) is the topological direct sum of H_1 and $V_3 = \mathrm{rg}(g)$ equipped with their respective subspace topologies (and where rg(g) denotes the range of g). Now we define the euclidean form $(\ ,\ )$ on V_1 in such a way that via $\mathrm{id} \oplus g$

$$(H_1,(\ ,\ )^1) \oplus (V_2,(\ ,\ )^2) \simeq (V_1,(\ ,\ )).$$

Then $(\ ,\ )$ induces μ_1 and H has the orthogonal complement V_3.

4.11 <u>Exercise</u>. Let T_{VC} be the theory T_C where the axioms "K is real closed with its order topology" are replaced by "K is V-topological". Show

a) Every locally bounded vector space over a complete field with an absolute value together with a closed subspace is a model of T_{VC}.

b) Two models $((K_i,\nu_i),((V_i,H_i),\mu_i))$ of T_{VC} are L_m-equivalent iff $K_1 \equiv K_2$, $\dim H_1 = \dim H_2$ and $\dim {}^{V_1}/H_1 = \dim {}^{V_2}/H_2$.

c) A structure is a model of T_{VC} iff it is L'_m-equivalent to a structure of the form

$$((K,\nu),((\bigoplus_{i\in I} K, \bigoplus_{i\in J} K),\mu)) \quad ,$$

where $J \subset I$ and where (K,ν) and μ are as in 4.8 c).

Now we look at dense subspaces. Let T_D be the L'_m-theory axiomatized by

$$T_R \quad \text{and} \quad \text{"P is a dense subspace"} .$$

4.12 <u>Theorem</u>. Two models $((K_i,\nu_i),((V_i,H_i),\mu_i))$, $i = 1,2$, of T_D are equivalent iff $\dim H_1 = \dim H_2$ and $\dim {}^{V_1}/H_1 = \dim {}^{V_2}/H_2$.

4.13 Remark. Let $((K,\nu),(V,\mu))$ be a euclidean topological vector space with a countable orthonormal basis $u_o, u_1, \dots$ (w.r.t. the given bilinear form). Suppose that the sequence $(a_i)_{i \in \omega}$ of scalars converges to $0 \in K$. Then, for $n \geq 1$, the set $\{u_o + a_i u_j \mid i \in \omega, i < j \leq i+n\}$ generates a dense subspace H with $\dim {}^V/H = n$. Similarly one obtains a dense subspace with infinite codimension.

Theorem 4.12 (as 4.1b) follows from 4.5 (and 4.6) and the following lemma.

4.14 Lemma. Let $(V_i, (\ ,\ ))$, $i = 1,2$, be two euclidean vector spaces over K of countable dimension. Suppose H_1 and H_2 are dense (w.r.t the topology induced by $(\ ,\ )$) subspaces with $\dim {}^{V_1}/H_1 = \dim {}^{V_2}/H_2$.
Then $(V_1, H_1, (\ ,\ ))$ and $(V_2, H_2, (\ ,\ ))$ are isomorphic.

Proof. We need the following fact about a dense subspace H in a euclidean vector space $(V, (\ ,\ ))$:

> If F is a finite dimensional subspace and $v \in V$,
> then $H \cap (v + F^{\perp})$ is dense in $v + F^{\perp}$.

Proof: Let $v_1, \dots v_n$ be an orthonormal basis of F. The orthogonal projection from V to F maps H onto F. Hence there are $u_1, \dots, u_n \in H$ s.t. $v_i - u_i \in F^{\perp}$, i.e. $(u_i, v_j) = (v_i, v_j) = \delta_{ij}$. Now suppose $a > 0$. Choose $g \in H$ with $\|v - g\| < a$. By Schwarz'inequality, $|\alpha_i| < a$ where $\alpha_i = (v_i, v - g)$. Set $h = g + \alpha_1 u_1 + \dots + \alpha_n u_n$. Then $h \in H \cap (v + F^{\perp})$ and $\|v - h\| < a(1 + \|u_1\| + \dots + \|u_n\|)$.

For the proof of 4.14 we assume first that $\dim {}^{V_i}/H_i = n < \infty$. Choose orthonormal $p_1, \dots, p_n \in V_1$ and $q_1, \dots, q_n \in V_2$ linearly independent modulo H_1 resp. H_2. We construct bases $u_1, u_2, \dots$ and $v_1, v_2, \dots$ of H_1 resp. H_2 s.t. for $i,j \geq 1$ and $r = 1, \dots, n$

(*) $$(u_i, p_r) = (v_i, q_r) \text{ and } (u_i, u_j) = (v_i, v_j)\ .$$

Then the linear map given by $u_i \mapsto v_i$ $(i \geq 1)$ and $p_r \mapsto q_r$ $(1 \leq r \leq n)$ will yield the desired isomorphism. The inductive definition of the elements of the bases uses the fact:

> Suppose $u_1, \dots, u_m \in H_1$ and $v_1, \dots, v_m \in H_2$ satisfy (*)
> and let $u_{m+1} \in H_1$. Then there is $v_{m+1} \in H_2$ s.t.

$$u_1,\dots,u_{m+1} \quad \text{and} \quad v_1,\dots,v_{m+1} \quad \text{satisfy} \quad (*).$$

To establish this, we set $b_i = (u_{m+1}, u_i)$, $c_r = (u_{m+1}, p_r)$ and

$$G_1 = \{x \in V_1 | (x,u_i) = b_i,\ (x,p_r) = c_r \quad \text{for } i = 1,\dots,m;\ r = 1,\dots,n\}$$

$$G_2 = \{x \in V_2 | (x,v_i) = b_i,\ (x,q_r) = c_r \quad \text{for } i = 1,\dots,m;\ r = 1,\dots,n\}\ .$$

We can assume that $u_{m+1} \notin \langle u_1,\dots,u_m\rangle$. Then $u_{m+1} \notin \langle u_1,\dots,u_m,p_1,\dots,p_n\rangle$ and hence u_{m+1} is not perpendicular to G_1. Thus, the distance from 0 to G_1 is smaller than $\|u_{m+1}\|$. This distance can be computed from the $b_i, c_r, (u_i,p_r)$ and (u_i,u_j). Hence by (*), G_2 has the same distance from 0. Since $H_2 \cap G_2$ is dense in G_2, we find $h \in H_2 \cap G_2$ s.t. $\|h\| \leq \|u_{m+1}\|$. But the dimension of the affine space $H_2 \cap G_2$ is > 0. Hence $H_2 \cap G_2$ also contains an element v_{m+1} with $\|v_{m+1}\| = \|u_{m+1}\|$.

In case that the dimension of V_i/H_i is infinite, we construct <u>simultaneously</u> four sequences

$$p_1,p_2,\dots\ ,\ u_1,u_2,\dots \quad \in V_1$$

$$q_1,q_2,\dots\ ,\ v_1,v_2,\dots \quad \in V_2$$

s.t. $u_1,u_2,\dots$ and $v_1,v_2,\dots$ form bases of H_1 resp. H_2, $p_1,p_2,\dots$ and $q_1,q_2,\dots$ are orthonormal,

$p_1 + H_1, p_2 + H_1,\dots$ and $q_1 + H_2, q_2 + H_2,\dots$ are bases of V_1/H_1 resp. V_2/H_2, and such that (*) holds.

This can be done using the following fact which is easy to prove:

> Let H be a subspace of V, $u_1,\dots,u_m \in H$ and $p_1,\dots,p_n \in V$.
> Given $p \in V$ there is $p_{n+1} \in V$ s.t.
> $\langle p_1,\dots,p_n,p\rangle + H = \langle p_1,\dots,p_n,p_{n+1}\rangle + H$ and
> p_{n+1} is orthogonal to $p_1,\dots,p_n,u_1,\dots,u_m$.

Again the desired isomorphism is given by $u_i \mapsto v_i$ $(i \geq 1)$ and $p_r \mapsto q_r$ $(r \geq 1)$.

The following theorem summarizes the preceding results.

Let T_S be the theory obtained from T_R adding the axiom "P is a subspace" and, for $n \in \omega$, the L'_m-sentence

(7) $\forall X\ \exists Y\ \forall x_1 \ldots \forall x_n\ \forall y \notin \bar{P} + \langle x_1,\ldots,x_n\rangle \exists x$

$$(x \in X \cap (\bar{P} + \langle x_1,\ldots,x_n,y\rangle) \wedge (x + Y) \cap (\bar{P} + \langle x_1,\ldots,x_n\rangle) = \emptyset)$$

(where $\bar{P}$ denotes the closure of P).

4.15 Theorem. a) Every locally bounded real vector space with a distinguished subspace is a model of T_S.

b) Two models $((K_i,\nu_i),((V_i,H_i),\mu_i))$, $i = 1,2$, of T_S are L'_m-equivalent iff

$\dim H_1 = \dim H_2$, $\dim {}^{\bar{H}_1}/H_1 = \dim {}^{\bar{H}_2}/H_2$ and $\dim {}^{V_1}/\bar{H}_1 = \dim {}^{V_2}/\bar{H}_2$.

Proof. Part a) follows from 4.9 a), since $((K,\nu),((V,H),\mu))$ is a model of T_S iff $((K,\nu),((V,\bar{H}),\mu))$ is a model of T_C.

One direction of b) is easy. Now assume that $((K_i,\nu_i),((V_i,H_i),\mu_i))$, $i = 1,2$, are models of T_S and have the "same dimensions". By 4.10 there are denumerable euclidean vector spaces $((K,\nu),((V_j,H_j),\mu_j))$, $j = 3,4$, s.t.

$$((K_i,\nu_i),((V_i,H_i),\mu_i)) \equiv_{L'_m} ((K,\nu),((V_{i+2},H_{i+2}),\mu_{i+2}))$$

and s.t. for $j = 3,4$, $\bar{H}_j$ has an orthogonal complement G_j (w.r.t the given form $(\ ,\ )_j$). Since $\dim G_3 = \dim G_4$, we have

$$(G_3,(\ ,\ )_3) \simeq_K (G_4,(\ ,\ )_4)$$

(where ... $\simeq_K$ --- means that the spaces ... and --- are K-isomorphic).

Since $\dim H_3 = \dim H_4$, $\dim {}^{\bar{H}_3}/H_3 = \dim {}^{\bar{H}_4}/H_4$ and H_j is dense in $\bar{H}_j$, we have by 4.14,

$$(\bar{H}_3,H_3,(\ ,\ )_3) \simeq_K (\bar{H}_4,H_4,(\ ,\ )_4)\ .$$

Putting corresponding isomorphisms together, we see that

$$(V_3,H_3,(\ ,\ )_3) \simeq_K (V_4,H_4,(\ ,\ )_4)\ .$$

Hence

$$((K\nu),((V_3,H_3),\mu_3)) \simeq ((K,\nu),((V_4,H_4),\mu_4))\ .$$

4.16 Corollary. T_S is decidable.

4.17 Remark. We do not know a version of 4.12 for V-topological fields.

C <u>Banach spaces with linear mappings</u>.

We look at structures of the form

$$((K,\nu),(V,\mu),(V^{+},\mu^{+}),f)$$

where (V,μ) and (V^{+},μ^{+}) are topological vector spaces over the same topological field (K,ν) and where $f:V \to V^{+}$ is a linear map. For the corresponding L'', let T_M be the L''_m-theory expressing that

(i) $((K,\nu),((V,\ker(f)),\mu) \models T_C$

(where $\ker(f)$ denotes the kernel of f).

(ii) $((K,\nu),((V^{+},rg(f)),\mu^{+})) \models T_C$

(iii) $f:V \to V^{+}$ is continuous and linear, and open as a map from V to $rg(f)$.

4.18 <u>Theorem</u>. a) Every continuous linear map between Banach spaces with closed range gives rise to a model of T_M.

b) Two models $((K_i,\nu_i),(V_i,\mu_i),(V_i^{+},\mu_i^{+}),f_i)$, $i = 1,2$, are L''_m-equivalent iff $\dim \ker(f_1) = \dim \ker(f_2)$, $\dim rg(f_1) = \dim rg(f_2)$ and $\dim V_1^{+}/_{rg(f_1)} = \dim V_2^{+}/_{rg(f_2)}$.

<u>Proof</u>. a) By the open mapping theorem any such map (as a map to its range) is open. Since the kernel is closed, the assertion follows from 4.9 a).

b) One direction is clear. For the other direction we argue as follows:

Let $((K,\nu),(V,\mu),(V^{+},\mu^{+}), f)$ be a model of T_M.

By 4.10 there is an L''_m-equivalent denumerable structure

$$((K_o,\nu_o),(V_o,\mu_o),(V_o^{+},\mu_o^{+}),f_o) ,$$

where μ_o and μ_o^{+} are induced by euclidean forms $(\ ,\ )$ and $(\ ,\ )^{+}$, and where $\ker(f_o)$ and $im(f_o)$ have orthogonal complements G and G^{+}.

Being open and continuous f yields a K-isomorphism of the topological vector spaces G and $rg(f_o)$. Whence, if $(\ ,\ )_o^{+}$ is the orthogonal sum of $(\ ,\ )^{+}{}_{|G^{+}}$ and the image of $(\ ,)_{|G}$, then $(\ ,\ )_o^{+}$ again induces μ_o^{+}.

But now a denumerable structure

$$(K,(V,G,(\ ,\ )),(V^{+},G^{+},(\ ,\ )^{+}),f) ,$$

where

$(V,(\ ,\ ))$ and $(V^+,(\ ,\ )^+)$ are euclidean vector spaces over K,

$f: V \to V^+$ is K-linear,

$V = G \oplus \ker(f)$, $V^+ = G^+ \oplus rg(f)$ (orthogonal direct sum)

$f_{|G}$ preserves $(\ ,\ )$,

is determined up to isomorphism by K, dim ker(f), dim rg(f) and dim G^+.

4.19 Remarks. a) An analogue result holds for continuous maps between "Banach spaces" over complete fields with an absolute value.

b) We have no results for continuous linear maps between Banach spaces without the assumption that the range is closed.

c) 4.18 says: All elementary properties of continuous, linear maps between Banach spaces with closed range can be elementarily derived from the Riesz' lemma and the open mapping theorem.

D. Dual pairs of normed spaces.

Let $(V,\|\ \|)$ be a real normed vector space. Denote by $(V',\|\|')$ the dual vector space with its canonical norm

$$\|f\| = \sup\{|f(u)| \mid u \in V, \|u\| = 1\}.$$

Let $[,]: V \times V' \to \mathbb{R}$ be the canonical bilinear form and μ and μ' the topologies induced by $\|\|$ resp. $\|\|'$.

We will show that the L^*_m-theory (for the corresponding L^*) of such a dual pair

$$((\mathbb{R},\varrho),(V,\mu),(V',\mu'),[,])$$

is determined by the dimension of V.

In the language L^*_m we use the following variables

$x_1, x_2, \ldots$ as variables for elements of V

$X_1, X_2, \ldots$ as variables for elements of μ

$y_1, y_2, \ldots$ as variables for elements of V'

$Y_1, Y_2, \ldots$ as variables for elements of μ'.

Let T_{DP} be an L_m^*- theory s.t. the models of T_{DP} are just the structures of the form

$$((K,\nu),(V,\mu),(V^+,\mu^+),[,]) ,$$

where $((K,\nu),(V,\mu)$ and $((K,\nu),(V^+,\mu^+))$ are models of T_R, $[,]:V \times V^+ \to K$ is bilinear and continuous and where the following axioms hold for $n = 0,1,2,\ldots$

(8) $\forall X\ \forall Y\ \exists A\ \forall x_1 \ldots \forall x_{n+1} (x_{n+1} \notin \langle x_1,\ldots,x_n\rangle$

$\to \exists x\ \exists y (x \in X \cap \langle x_1,\ldots,x_{n+1}\rangle \wedge [x,y] = 1 \wedge [x_1,y] = \ldots = [x_n,y] = 0 \wedge Ay \subset Y)$

(9) $\forall Y\ \forall X\ \exists A\ \forall y_1 \ldots \forall y_{n+1} (y_{n+1} \notin \langle y_1,\ldots,y_n\rangle$

$\to \exists y\ \exists x (y \in Y \cap \langle y_1,\ldots,y_{n+1}\rangle \wedge [x,y] = 1 \wedge [x,y_1] = \ldots = [x,y_n] = 0 \wedge Ax \subset X)$.

4.20 Theorem. a) Every dual pair belonging to a real normed vector space is a model of T_{DP}.

b) Two models $((K_i,\nu_i),(V_i,\mu_i),(V_i^+,\mu_i^+),[,]_i)$, $i = 1,2$, of T_{DP} are L_m^*-equivalent iff $\dim V_1 = \dim V_2$.

Proof. a) Let $(V,\|\|)$ be a real normed vector space. To prove (8), suppose that w.l.o.g. $B_a = \{x \mid \|x\| \leq a\}$ and $B'_a = \{y \mid \|y\|' \leq a\}$ are given for X resp. Y. Take as A the set $(-\frac{a^2}{2},\frac{a^2}{2})$. We show that it satisfies (8). Let $u_{n+1} \notin \langle u_1,\ldots,u_n\rangle$ be given. Choose $u \in \langle u_1,\ldots,u_{n+1}\rangle$ by Riesz' lemma s.t. $u \in B_a$ and

$(u + B_{\frac{a}{2}}) \cap \langle u_1,\ldots,u_n\rangle = \emptyset$. Then the linear functional $g:\langle u_1,\ldots,u_{n+1}\rangle \to \mathbb{R}$ with $g(u_1) = \ldots = g(u_n) = 0$, $g(u) = 1$ has a norm $\leq \frac{2}{a}$. The Hahn-Banach theorem yields $v \in V'$ s.t. $[u_1,v] = \ldots = [u_n,v] = 0$, $[u,v] = 1$ and $\|v\|' \leq \frac{2}{a}$. We have $(-\frac{a^2}{2},\frac{a^2}{2})v \subset B'_a$.

For the proof of (9) we proceed similarly. Let X,Y and A be as above and suppose $v_1,\ldots,v_{n+1} \in V'$, $v_{n+1} \notin \langle v_1,\ldots,v_n\rangle$ are given. By Riesz' lemma we get $v \in \langle v_1,\ldots,v_n\rangle$ s.t. $v \in B_a$ and $(v + B_{\frac{2}{3}a}) \cap \langle v_1,\ldots,v_n\rangle = 0$. Now Helly's theorem states:

Given $a_1,\dots,a_{n+1},b,c \in \mathbb{R}, b,c > 0$ and $\bar{v}_1,\dots,\bar{v}_{n+1} \in V'$ with

$$|b_1a_1+\dots+b_{n+1}a_{n+1}| \le b\|b_1\bar{v}_1+\dots+b_{n+1}\bar{v}_{n+1}\|' \quad \text{for all } b_1,\dots,b_{n+1} \in \mathbb{R},$$

there is $u \in V$ s.t. $[u,\bar{v}_i] = a_i$ $(i = 1,\dots,n+1)$ and $\|u\| \le b + c$ (cf. [27] p. 109. The proof given there works for arbitrary normed spaces).

Put $a_1=\dots=a_n = 0$, $a_{n+1} = 1$, $b = \frac{3}{2a}$ and $c = \frac{1}{2a}$ and $\bar{v}_1 = v_1,\dots,\bar{v}_n = v_n$ and $\bar{v}_{n+1} = v$.

Then we can apply Helly's theorem and obtain $u \in V$ s.t.

$$[u,v_1]=\dots=[u,v_n] = 0, \ [u,v] = 1 \ \text{ and } \ \|u\| \le \frac{2}{a} .$$

Part b) of the theorem will follow immediately from the lemma:

4.21 Lemma. A structure is a model of T_{DP} iff it is L^*_m-equivalent to a (denumerable) structure $((K,\nu),(V,\mu),(V^+,\mu^+),[,])$, where $((K,\nu),(V,\mu))$ and $((K,\nu,),(V^+,\mu^+))$ are euclidean topological vector spaces with euclidean forms $(\ ,\ )$ resp. $(\ ,\ )^+$, and where $[,]$ is defined by

$$[\Sigma a_i\, \bar{u}_i, \Sigma b_i\, \bar{v}_i] = \Sigma a_i b_i$$

for suitable orthonormal bases $\bar{u}_1,\bar{u}_2,\dots$ and $\bar{v}_1,\bar{v}_2,\dots$ of V resp. V^+.

Proof. First let $((K,\nu),(V,\mu),(V^+,\mu^+),[,])$ be as above. $[,]$ is continuous since $|[u,v]| \le \|u\|\|v\|^+$.

It is enough to prove (8). Take as X,Y and A the sets B_a, B_a^+ resp. $(-a^2,a^2)$. Given $u_1,\dots,u_{n+1} \in V$, $u_{n+1} \notin \langle u_1,\dots,u_n\rangle$ choose $u \in \langle u_1,\dots,u_{n+1}\rangle$ orthogonal to $\langle u_1,\dots,u_n\rangle$ and of length a. Suppose $u = \Sigma a_i\bar{u}_i$. Set $v = \frac{1}{a^2}(\Sigma a_i\bar{v}_i)$. Then $[u_1,v]=\dots=[u_n,v] = 0$, $[u,v] = 1$, $\|v\|^+ = \frac{1}{a}$ and $(-a^2,a^2)v \in B_a^+$.

Now suppose that $((K,\nu),(V,\mu),(V^+,\mu^+))$ is an – ω_1-saturated – model of T_{DP}. Let $W \in \nu$ be bounded. Choose bounded $U \in \mu$ and $U^+ \in \mu^+$ s.t.

$$[u,v] \in W \ \text{ for all } \ u \in U,\ v \in U^+ .$$

There is $W_1 \in \nu$ small enough s.t.

for all $u_{n+1} \notin \langle u_1,\dots,u_n\rangle$ there are $u \in U \cap \langle u_1,\dots,u_{n+1}\rangle$ and $v \in V^+$ s.t. $[u_1,v]=\dots=[u_n,v] = 0, [u,v] = 1$ and $W_1 v \in U^+$,

and

for all $v_{n+1} \notin \langle v_1,\dots,v_n\rangle$ there are $v \in U^+ \cap \langle v_1,\dots,v_{n+1}\rangle$ and $u \in V$ s.t. $[u,v_1]=\dots=[u,v_n] = 0, [u,v] = 1$ and $W_1 v \in U$.

Pick $a \in W_1 \setminus \{0\}$ and choose a valuation ring $A \subset K$ s.t. $\nu = \nu_{A}, \frac{1}{a} \in A$ and $W \subset A$. Let D (resp. D^+) be the A-module generated by U (resp. U^+). Since μ and μ^+ are closed under countable intersections, D and D^+ are bounded and we have

$$[u,v] \in A \quad \text{for all } u \in D,\ v \in D^+,$$

for all $u_{n+1} \notin \langle u_1,\dots,u_n\rangle$ there are $u \in D \cap \langle u_1,\dots,u_{n+1}\rangle$ and $v \in D^+$ s.t. $[u_1,v]=\dots=[u_n,v] = 0, [u,v] = 1$,

and

for all $v_{n+1} \notin \langle v_1,\dots,v_n\rangle$ there are $v \in D^+ \cap \langle v_1,\dots v_{n+1}\rangle$ and $u \in D$ s.t. $[u,v_1]=\dots=[u,v_n] = 0, [u,v] = 1$.

Now we proceed to a denumerable structure L''_m-equivalent to $(((K,A),\nu),((V,D),\mu),((V^+,D^+),\mu^+))$, which we denote in the same way.

We are going to construct two bases $u_1,u_2,\dots$ and $v_1,v_2,\dots$ of V resp. V^+ s.t.

(*) $\quad u_i \in D,\ v_i \in D^+,\ [u_i,v_j] = \delta_{ij}$.

Let $u_1,\dots,u_n \in V$ and $v_1,\dots,v_n \in V^+$ be given s.t. (*) holds. Given $\bar{u} \in V$ (or $\bar{v} \in V$), we have to extend the sequences in such a way that $\bar{u} \in \langle u_1,\dots,u_{n+1}\rangle$ (or $\bar{v} \in \langle v_1,\dots,v_{n+1}\rangle$). We only look at the case $\bar{u} \in V$ (since the situation is symmetric).

There is nothing to be proved if $\bar{u} \in \langle u_1,\dots,u_n\rangle$.

Thus, we assume $\bar{u} \notin \langle u_1,\dots,u_n\rangle$. Then there is $u \in D \cap \langle u_1,\dots,u_n,u\rangle$ and $v_{n+1} \in D^+$ s.t.

$$[u_1,v_{n+1}]=\dots=[u_n,v_{n+1}] = 0,\ [u,v_{n+1}] = 1.$$

We set $u_{n+1} = u - ([u,v_1]u_1+\ldots+[u,v_n]u_n)$. Then (*) holds and $\bar{u} \in \langle u_1,\ldots,u_{n+1}\rangle$.

Choose the euclidean forms $(\ ,\ )$ and $(\ ,\ )^+$ s.t. $u_1,u_2,\ldots$ resp. $v_1,v_2,\ldots$ are orthonormal. It remains to show that the topologies induced by these forms, are μ resp. μ^+.

First note that we have $[-b,b] \subset A \subset [-c,c]$ for some $b,c > 0$. Given $d > 0$, we have $\frac{d}{c} D \subset B_d$. For $\Sigma a_i u_i \in D$ implies $a_j = [\Sigma a_i u_i, v_j] \in A$ and $\sqrt{\Sigma a_i^2} \in A$. Conversely, given $U_1 \in \mu$ choose $e > 0$ s.t. $eD \subset U_1$. Then $B_{be} \subset U_1$, since $\sqrt{\Sigma a_i^2} \leq be$ implies $|a_i| \leq be$ and $\Sigma a_i x_i \in b\,e\,D$.

4.22 Remarks. a) 4.21 can be generalized to arbitrary V-topological fields (see 4.8). But it is not clear what is the generalization of 4.20a.

b) Note that in models of T_{DP}, μ^+ is uniquely determined by μ. This is a special case of I.8.8.6 since

$$\forall Y\ \forall X\ \exists A\ \forall y((\forall x \in X[x,y] \in A) \rightarrow y \in Y)$$

$$\exists X\ \forall A\ \exists Y\ \forall y(y \in Y \rightarrow \forall x \in X[x,y] \in A)$$

follow from T_{DP}.

It is easy to see that μ^+ is not explicitly definable from μ (cf. I.7.6).

Historical remarks

§ 1 The results of this section are due to the second author. 1.23 b can be derived from [14]. Theorem 1.24 also follows from [10].
1.9 (with a similar proof) was independently found by L. Heindorf, who also has proved some results on decidable non-T_3 spaces. 1.50 e) is due to Heindorf. [1] contains categoricity results for L_t+ weak monadic second order quantifiers. 1.55 d) is due to J. Strobel, who determines the L_m-elementary theories of a lot of uniform spaces and proximity spaces.

§ 2 2.8 and 2.14 are first proved in [6]. The proofs given in this book and 2.11 are due to the second author.

§ 3 The results of this section are taken from [15]. V-topological fields were introduced in [11]. 3.8 was first proved in [25] (by a related method).

§ 4 All theorems are taken from [24], which originated in work of the second author. 4.14 can be derived from [28]. The given proofs and the axiomatization of T_{DP} are due to the second author.

References

[1] G. Ahlbrand: Endlich axiomatisierbare Theorien von T_3-Räumen, Diplomarbeit, Freiburg (1979).

[2] W. Baur: Undecidability of the theory of abelian groups with a subgroup, Proc. AMS 55 (1976), pp. 125-128.

[3] N. Bourbaki: Algèbre (Modules sur les anneaux principaux), Paris (1964).

[4] N. Bourbaki: Algèbre commutative (Valuations), Paris (1964).

[5] Charlotte N. Burger: Some remarks on countable topological spaces. Seminarreport, FU Berlin (1971).

[6] G. Cherlin, P. Schmitt: Decidability of topological abelian groups, (1979) to appear.

[7] P.C. Eklof, E.R. Fischer: The elementary theory of abelian groups, Annals math. logic 4 (1972), pp. 115-171.

[8] L. Fuchs: Infinite abelian groups, Vol. I, New York (1970).

[9] Y. Gurevich: Expanded theory of ordered abelian groups, Annals of math. logic 12 (1977), pp. 193-228.

[10] Y. Gurevich: Monadic theory of order and topology, Israel J. of Math. 27 (1977), pp. 299-319.

[11] I. Kaplansky: Topological methods in valuation theory, Duke Math.J. 14 (1947), pp. 527-541.

[12] H.J. Kowalsky, H. Dürbaum: Arithmetische Kennzeichnung von Körpertopologien, J. reine angew. Math. 191 (1953), pp. 135-152.

[13] H. Läuchli, J. Leonhard: On the elementary theory of linear order, Fund. Math. 59, pp. 109-116.

[14] I.L. Lynn: Linearly orderable spaces, Trans. AMS 113 (1964), pp. 189-218.

[15] A. Prestel, M. Ziegler: Model-theoretic methods in the theory of topological fields, J. reine angewandte Math. 299/300 (1978), pp. 318-341.

[16] A. Prestel, M. Ziegler: Non axiomatizable classes of V-topological fields, to appear.

[17] M.O. Rabin: Decidability of second-order theories and automata on infinite trees, Trans. AMS. 141, (1969), pp. 1-35.

[18] M.O. Rabin: A simple method for undecidability proofs and some applications, in Bar-Hillel (Ed.) Logic, Meth. and Phil. (1965), pp. 58-68.

[19] A. Robinson: Complete theories, Amsterdam (1956).

[20] G.E. Sacks: Saturated Model Theory, Reading (1972).

[21] W.R. Scott: Algebraically closed groups, Proc. Amer. Math. Soc. 2 (1951), pp. 118-121.

[22] D. Seese: Decidability of ω-trees with bounded sets, in print.

[23] W. Szmielew: Elementary properties of abelian groups, Fund. Math. 41 (1955), pp. 203-271.

[24] V. Sperschneider: Modelltheorie topologischer Vektorräume, Dissertation Freburg, in preparation.

[25] A.L. Stone: Nonstandard analysis in topological algebra, in Applications of Model Theory to Algebra, Analysis and Probability, New York (1969), pp. 285-300.

[26] J.P. Thomas: Associted regular spaces, Canadian Journal 20, (1968), pp. 1087-1092.

[27] K. Yosida: Functional Analysis, Berlin (1964).

[28] H. Gross: Eine Bemerkung zu dichten Unterräumen reeller quadratischer Räume, Comm. Math. Helv. 45 (1970), pp. 472-493.

Subject index

Index of symbols

E r r a t a

page	line	read	for
8	19	...regular and Hausdorff...	... regular ...
15	22	$\hat{p} = (\ldots$	$p = (\ldots$
18	6	...$p = (\{(a_i, \ldots$	... $p = (\{a_i, \ldots$
23	26	...with $\models_t \varphi \leftrightarrow \psi$.	... with $\models \varphi \leftrightarrow \psi$.
28	11	...be universal and positive...	... be positive...
39	19	...any $\varphi \in L_t$	... any $\psi \in L_t$
40	3	$\varphi^{\mathfrak{A}} = \{\ldots$	$\varphi = \{\ldots$
40	5	...$\sigma = \varphi^{\widetilde{\mathfrak{A}}}$...	... $\sigma = \varphi^{\tilde{\mathfrak{A}}}$
57	12	...$\forall X\ \exists y\ \forall x(\neg Xxy$...	... $\forall X\ \exists y\ \forall x(\neg Uxy$...
60	last	$(\mathfrak{A},\delta) \models \psi^p$ ...	$(\mathfrak{A},\delta) \models \varphi^p$ ...
65	11	...$\mathfrak{B}^* \models Icx\varphi(x)$...	... $\mathfrak{B}^* \models Icx\psi(x)$...
68	last	then $\Theta_n = \delta_n$.	then $\otimes_n = \delta_n$.
71	3	...over denumerable models...	... over models...
90	3	...successors of a.	... successors of b.
91	11	...$U_X(t,s)$...	... $U_X(t)$...
91	12	...$U_X(t,s)$...	... $U_X(t)$...
143	26	...$u \in D \cap \langle u_1, \ldots, u_n, \bar{u}\rangle$...	... $u \in D \cap \langle u_1, \ldots, u_n, u\rangle$...
50	20	...each satisfiable denumerable	... each denumerable

The reference for the work of Heindorf (compare "Historical remarks" p.145) is :

Heindorf, L.: <u>Entscheidungsprobleme topologischer Räume</u>, Humboldt Universität Berlin (1979).

Vol. 609: General Topology and Its Relations to Modern Analysis and Algebra IV. Proceedings 1976. Edited by J. Novák. XVIII, 225 pages. 1977.

Vol. 610: G. Jensen, Higher Order Contact of Submanifolds of Homogeneous Spaces. XII, 154 pages. 1977.

Vol. 611: M. Makkai and G. E. Reyes, First Order Categorical Logic. VIII, 301 pages. 1977.

Vol. 612: E. M. Kleinberg, Infinitary Combinatorics and the Axiom of Determinateness. VIII, 150 pages. 1977.

Vol. 613: E. Behrends et al., L^P-Structure in Real Banach Spaces. X, 108 pages. 1977.

Vol. 614: H. Yanagihara, Theory of Hopf Algebras Attached to Group Schemes. VIII, 308 pages. 1977.

Vol. 615: Turbulence Seminar, Proceedings 1976/77. Edited by P. Bernard and T. Ratiu. VI, 155 pages. 1977.

Vol. 616: Abelian Group Theory, 2nd New Mexico State University Conference, 1976. Proceedings. Edited by D. Arnold, R. Hunter and E. Walker. X, 423 pages. 1977.

Vol. 617: K. J. Devlin, The Axiom of Constructibility: A Guide for the Mathematician. VIII, 96 pages. 1977.

Vol. 618: I. I. Hirschman, Jr. and D. E. Hughes, Extreme Eigen Values of Toeplitz Operators. VI, 145 pages. 1977.

Vol. 619: Set Theory and Hierarchy Theory V, Bierutowice 1976. Edited by A. Lachlan, M. Srebrny, and A. Zarach. VIII, 358 pages. 1977.

Vol. 620: H. Popp, Moduli Theory and Classification Theory of Algebraic Varieties. VIII, 189 pages. 1977.

Vol. 621: Kauffman et al., The Deficiency Index Problem. VI, 112 pages. 1977.

Vol. 622: Combinatorial Mathematics V, Melbourne 1976. Proceedings. Edited by C. Little. VIII, 213 pages. 1977.

Vol. 623: I. Erdelyi and R. Lange, Spectral Decompositions on Banach Spaces. VIII, 122 pages. 1977.

Vol. 624: Y. Guivarc'h et al., Marches Aléatoires sur les Groupes de Lie. VIII, 292 pages. 1977.

Vol. 625: J. P. Alexander et al., Odd Order Group Actions and Witt Classification of Innerproducts. IV, 202 pages. 1977.

Vol. 626: Number Theory Day, New York 1976. Proceedings. Edited by M. B. Nathanson. VI, 241 pages. 1977.

Vol. 627: Modular Functions of One Variable VI, Bonn 1976. Proceedings. Edited by J.-P. Serre and D. B. Zagier. VI, 339 pages. 1977.

Vol. 628: H. J. Baues, Obstruction Theory on the Homotopy Classification of Maps. XII, 387 pages. 1977.

Vol. 629: W. A. Coppel, Dichotomies in Stability Theory. VI, 98 pages. 1978.

Vol. 630: Numerical Analysis, Proceedings, Biennial Conference, Dundee 1977. Edited by G. A. Watson. XII, 199 pages. 1978.

Vol. 631: Numerical Treatment of Differential Equations. Proceedings 1976. Edited by R. Bulirsch, R. D. Grigorieff, and J. Schröder. X, 219 pages. 1978.

Vol. 632: J.-F. Boutot, Schéma de Picard Local. X, 165 pages. 1978.

Vol. 633: N. R. Coleff and M. E. Herrera, Les Courants Résiduels Associés à une Forme Méromorphe. X, 211 pages. 1978.

Vol. 634: H. Kurke et al., Die Approximationseigenschaft lokaler Ringe. IV, 204 Seiten. 1978.

Vol. 635: T. Y. Lam, Serre's Conjecture. XVI, 227 pages. 1978.

Vol. 636: Journées de Statistique des Processus Stochastiques, Grenoble 1977, Proceedings. Edité par Didier Dacunha-Castelle et Bernard Van Cutsem. VII, 202 pages. 1978.

Vol. 637: W. B. Jurkat, Meromorphe Differentialgleichungen. VII, 194 Seiten. 1978.

Vol. 638: P. Shanahan, The Atiyah-Singer Index Theorem, An Introduction. V, 224 pages. 1978.

Vol. 639: N. Adasch et al., Topological Vector Spaces. V, 125 pages. 1978.

Vol. 640: J. L. Dupont, Curvature and Characteristic Classes. X, 175 pages. 1978.

Vol. 641: Séminaire d'Algèbre Paul Dubreil, Proceedings Paris 1976–1977. Edité par M. P. Malliavin. IV, 367 pages. 1978.

Vol. 642: Theory and Applications of Graphs, Proceedings, Michigan 1976. Edited by Y. Alavi and D. R. Lick. XIV, 635 pages. 1978.

Vol. 643: M. Davis, Multiaxial Actions on Manifolds. VI, 141 pages. 1978.

Vol. 644: Vector Space Measures and Applications I, Proceedings 1977. Edited by R. M. Aron and S. Dineen. VIII, 451 pages. 1978.

Vol. 645: Vector Space Measures and Applications II, Proceedings 1977. Edited by R. M. Aron and S. Dineen. VIII, 218 pages. 1978.

Vol. 646: O. Tammi, Extremum Problems for Bounded Univalent Functions. VIII, 313 pages. 1978.

Vol. 647: L. J. Ratliff, Jr., Chain Conjectures in Ring Theory. VIII, 133 pages. 1978.

Vol. 648: Nonlinear Partial Differential Equations and Applications, Proceedings, Indiana 1976–1977. Edited by J. M. Chadam. VI, 206 pages. 1978.

Vol. 649: Séminaire de Probabilités XII, Proceedings, Strasbourg, 1976–1977. Edité par C. Dellacherie, P. A. Meyer et M. Weil. VIII, 805 pages. 1978.

Vol. 650: C*-Algebras and Applications to Physics. Proceedings 1977. Edited by H. Araki and R. V. Kadison. V, 192 pages. 1978.

Vol. 651: P. W. Michor, Functors and Categories of Banach Spaces. VI, 99 pages. 1978.

Vol. 652: Differential Topology, Foliations and Gelfand-Fuks-Cohomology, Proceedings 1976. Edited by P. A. Schweitzer. XIV, 252 pages. 1978.

Vol. 653: Locally Interacting Systems and Their Application in Biology. Proceedings, 1976. Edited by R. L. Dobrushin, V. I. Kryukov and A. L. Toom. XI, 202 pages. 1978.

Vol. 654: J. P. Buhler, Icosahedral Golois Representations. III, 143 pages. 1978.

Vol. 655: R. Baeza, Quadratic Forms Over Semilocal Rings. VI, 199 pages. 1978.

Vol. 656: Probability Theory on Vector Spaces. Proceedings, 1977. Edited by A. Weron. VIII, 274 pages. 1978.

Vol. 657: Geometric Applications of Homotopy Theory I, Proceedings 1977. Edited by M. G. Barratt and M. E. Mahowald. VIII, 459 pages. 1978.

Vol. 658: Geometric Applications of Homotopy Theory II, Proceedings 1977. Edited by M. G. Barratt and M. E. Mahowald. VIII, 487 pages. 1978.

Vol. 659: Bruckner, Differentiation of Real Functions. X, 247 pages. 1978.

Vol. 660: Equations aux Dérivée Partielles. Proceedings, 1977. Edité par Pham The Lai. VI, 216 pages. 1978.

Vol. 661: P. T. Johnstone, R. Paré, R. D. Rosebrugh, D. Schumacher, R. J. Wood, and G. C. Wraith, Indexed Categories and Their Applications. VII, 260 pages. 1978.

Vol. 662: Akin, The Metric Theory of Banach Manifolds. XIX, 306 pages. 1978.

Vol. 663: J. F. Berglund, H. D. Junghenn, P. Milnes, Compact Right Topological Semigroups and Generalizations of Almost Periodicity. X, 243 pages. 1978.

Vol. 664: Algebraic and Geometric Topology, Proceedings, 1977. Edited by K. C. Millett. XI, 240 pages. 1978.

Vol. 665: Journées d'Analyse Non Linéaire. Proceedings, 1977. Edité par P. Bénilan et J. Robert. VIII, 256 pages. 1978.

Vol. 666: B. Beauzamy, Espaces d'Interpolation Réels: Topologie et Géometrie. X, 104 pages. 1978.

Vol. 667: J. Gilewicz, Approximants de Padé. XIV, 511 pages. 1978.

Vol. 668: The Structure of Attractors in Dynamical Systems. Proceedings, 1977. Edited by J. C. Martin, N. G. Markley and W. Perrizo. VI, 264 pages. 1978.

Vol. 669: Higher Set Theory. Proceedings, 1977. Edited by G. H. Müller and D. S. Scott. XII, 476 pages. 1978.